FALLING FOR GRAVITY

FALLING FOR GRAVITY

Invisible Forces in Contemporary Art

Catherine James

PETER LANG

Oxford • Bern • Berlin • Bruxelles • Frankfurt am Main • New York • Wien

Bibliographic information published by Die Deutsche Nationalbibliothek.
Die Deutsche Nationalbibliothek lists this publication in the Deutsche National-
bibliografie; detailed bibliographic data is available on the Internet at http://dnb.d-nb.de.

A catalogue record for this book is available from the British Library.

Library of Congress Cataloging-in-Publication Data

Names: James, Catherine, 1962- author.
Title: Falling for gravity : invisible forces in contemporary art / Catherine
  James.
Description: Oxford ; New York : Peter Lang, 2018. | Includes bibliographical
  references and index.
Identifiers: LCCN 2017028844 | ISBN 9783034317269 (alk. paper)
Subjects: LCSH: Gravity in art. | Art, Modern--21st century--Themes, motives.
Classification: LCC N8217.G747 J36 2017 | DDC 709.05--dc23 LC record available at
  https://lccn.loc.gov/2017028844

Cover image: Stan Douglas, *Intrigue, 1948*, 2010. Digital fibre print mounted on Dibond aluminium. 45.7 × 45.7 cm (18 × 18 in.). Courtesy of the artist, David Zwirner, New York/ London and Victoria Miro, London. © Stan Douglas.

Cover design by Peter Lang Ltd.

ISBN 978-3-0343-1726-9 (print) • ISBN 978-1-78707-820-8 (ePDF)
ISBN 978-1-78707-821-5 (ePub) • ISBN 978-1-78707-822-2 (mobi)

© Peter Lang AG, International Academic Publishers, Bern 2018
Wabernstrasse 40, CH-3007 Bern, Switzerland
info@peterlang.com, www.peterlang.com, www.peterlang.net

This publication has been peer reviewed.

Printed in Germany

For Susan and Graham

Contents

Illustrations

Acknowledgements

I must first thank my parents, Susan and Graham James, for their love, support, and encouragement during the book's formation. Without them, this project would never have reached completion. I would also like to convey particular thanks to Dr Kathy Battista, Programme Director for Contemporary Art at Sotheby's, New York, who has provided so many valuable references and connections to artworks and contemporary artists over the years, enriching the research process incalculably. Special thanks go to Victoria Miguel, whose editorial advice and expertise helped enormously during the closing stages of the book, in addition to the excellent editorial suggestions made by Dr Barbara Penner (Bartlett, UCL).

When I first started thinking about the role of gravity within art and culture as part of my doctoral research at the London Consortium in the late 1990s, I felt quite alone. Nevertheless, I was encouraged by Professor Steven Connor and the late Professor Paul Hirst to pursue this strange field of enquiry. Since my thesis was completed in 2004, a few invitations have arrived for me to speak about my ideas. I would like to thank Dr Anna Dezeuze for her invitation to speak at an international conference *Dwelling, Walking, Falling*, hosted at University of Manchester in 2009. I would also like to mention Emilyn Claid and Ric Allsopp, who included my article in a special issue, *On Falling*, in the journal *Performance Research* (2013). Thanks are due to Dr Davide Deriu (University of Westminster) for an invitation to participate in the symposium *Vertigo in the City: Conversations between the Sciences, Arts and Humanities* in 2014. These fascinating conversations and experiences have all helped me to feel a little less alone through the research process. Sincere gratitude goes to my former friends and colleagues at the London Consortium for their many insights over the years: Dr Mark Morris, Dr John Tercier, Dr Ruth Adams, Dr Kathy Battista, and Dr Bernard Vere.

Research for this book was made possible by an award from University of the Arts in London and, in its early stages, through my post as Lecturer in Modern & Contemporary Art at Christie's Education. Being able to travel to museums, galleries and art world events across Europe and to work alongside Dr Mike Ricketts and Dr Peter Higginson stimulated and enriched my research immeasurably. I am also grateful to the many artists and gallery archivists, who have assisted with the image research. Thanks also go to my friends and family for their loving support: Joshua James, Brendan and Suzanne James, Ererton Harrison, Michael Kemp, Charlotte May, Leeann Owens, Bobby Demosthenis, April Gilbin, and Charlie Coomes.

Introduction

I fell for gravity sometime in the late 1990s, whilst looking at an object of contemporary art that seemed to hover mysteriously in its own space without wires, strings, or support. I was left with the impression that this enigmatic floating object had more to answer for and, over time, I noticed how frequently many artists, from the 1960s onwards, stir something in their work by making materials, objects or bodies float, fly, lean, or collapse through the membrane of gravity. From Yves Klein's ecstatic leap into the void of 1960, through Bas Jan Ader's anti-heroic falling works of the early 1970s to the collapsing objects of Fischli and Weiss's film, *The Way Things Go*, of the late 1980s and the tightrope walk in Catherine Yass's *High Wire* of 2008, a subtle trope seems to hover across the field of contemporary art. If, as I suggest, gravity is more consciously arbitrated by contemporary art, sculpture, and performance, then what model of gravity is envisaged?

Gravity is a strange word and even stranger phenomenon, arousing both metaphysical wonder and feet-on-the-ground reality. Arbitrating both manifestations, Newton later popularized his unified theory of gravity in the *Principia* (1687) through the tale of an apple that fell on his head, whilst sitting in his garden at Woolsthorpe Manor. Relating this story to William Stukeley later in life, Newton's apocryphal apple was said to have launched his research into a theory of gravity that equated falling bodies on earth with planetary motion in the heavens. Combining the celestial physics of Kepler with the ground-based physics of Galileo, Newton set out an explanation of universal attraction, freed from the rationale of mechanical contact. Newton's discovery was seismic for both science and society. For the first time, artisans could predict and calculate mechanical forces accurately and, as a result, terrestrial mechanics was institutionalized, laying the foundation for the Industrial Revolution and its vast engineering projects. However, in the *Principia*, Newton did not discover a mechanical cause for gravity itself so that, ultimately, it remained occult, and potentially divine in origin.

It was not until the early 1900s that Albert Einstein started to muddy the waters, questioning Newton's reliance on mass as the force that kept things in place. Einstein's achievement was to recognize that gravity was a fundamental effect within the warp and weft of a 'space-time' continuum. As Robert Trotta explains, 'For Mr Newton, space and time did not talk to each other, never married and lived different lives.'[1] Instead of the falling apple, Einstein was to imagine a falling man, recognizing that the man does not feel his weight as he plunges downwards. From the position of the workman on top of a roof, if he falls with his toolbox, then both fall relative to each other and at the same speed. But the simple question Einstein posed was whether the fall was due to the pull of gravity or to the acceleration of the earth upwards. What he proved in his famous 'lift in space' thought experiment was that gravity and acceleration are equivalent. If a man floats in a lift within a weightless context of deep space, and, if then, an accelerating force is applied to the lift, he would stop floating and regain his normal gravitational upright stance within the lift. Without windows or external reference points, he could imagine himself standing on the earth. Einstein's rejection of Newton's reading of gravity was based on questioning how the frame of reference for gravity works, since the effects of stability are so powerfully present in our lives. A table sits on the floor in a building set on foundations based on the ground; however, Einstein destroyed these practical certainties. Instead gravity was theorized as a function or effect of curved space-time, thus unifying his relativity theory with the gravitational field. As a result, space and time were married forever. In a more radical statement, the contemporary American physicist Michio Kaku claims that gravity does not exist and that it is essentially the distortion of space-time that moves the stars and planets.[2] If gravity remains simply effect without cause, then its significance might be better explained as one of vanishing mediation. We follow its trail and effects, but can never really seize it.

---

1   Roberto Trotta, *The Edge of Sky: All you need to know about the all there is* (New York: Basic Books, 2014), 16.
2   Marcus Chown, *The Ascent of Gravity* (London: Weidenfeld and Nicholson, 2017), 129.

Since gravity is still being theorized, and is viewed as the fundamental pathway to understanding the universe, then these remarks remain contingent on what comes next. With recent discoveries of the Higgs boson particle and gravitational waves, the science of gravity is continually on the move. However, the artistic iterations of gravity are often refractory, still lodged in the Newtonian world of clockwork time and ordered space, or just reaching towards Einstein's space-time co-efficient in Relativity Theory. Therefore, those artists who explicitly work with gravity often seem to appropriate the simple empiricism of early experiments in physics. In this sense, they still intuit gravity through a persistent Newtonian viewpoint, relying on an inertial frame of reference to make sense of it. Furthermore, this quality of the experiment is closely linked to the idea of experience as grounded in the body. In other examples, there is a residual resistance to explanation, with artists creating play out of the occult quality of this invisible force. Newton's fascination with the occult was to preserve divine cause; however, in other religious readings, the material world is not considered part of a divine plan, but issues from the malign actions of a demiurge. In the Gnostic tradition, the world merely consisted of gross matter that imprisoned the human spirit and kept it distant from God, truth, and enlightenment. In fact, the Greek etymology of 'demiurge' is 'artisan' or 'craftsman' and it is this facet of meaning that resonates with the kind of art that is increasingly associated with contemporary art practice, as I hope to show.

Gravity, particularly in its post-Newtonian variant, is naturally claimed within scientific discourse and lies beyond the understanding of the non-specialist. This has naturally stymied its adoption within the arts. Using the word, 'gravity' within art and culture seems somehow audacious, though its popularized definition has started to appear more regularly in books and television documentaries. Yet, this force is the very foundation of life, the medium of all bodies and objects on earth, defining our size, height and weight, as well as our daily exertions, rituals and movements. Theoretical physicist Kip Thorne has even posited the strange idea that objects like to live in a region where they can exist the longest. Therefore, since gravity dilates time, it becomes a mysterious co-efficient determining how long we live. There may be grounds to suggest that this dilation of time is a critical quality of how artists approach matter; by particularizing and magnifying gravity's effect on matter; they dwell in materials, producing a stay against the passage of time.

As an amateur, I can only tease out the language of gravity in rudimentary fashion. Crudely put, mass defines the actual amount of matter in any given object, and weight is the effect of gravity on mass. This means that in a changed gravitational field, the weight of a given mass will change. The word, 'mass' does not just pertain to clumps of matter, but is contiguous with other meanings such as accumulation, aggregation, and combination. Mass is also a way of denoting the common people, the crowd or multitude, evading the nuances of individual character or differentiation. By contrast, artists particularize things, give a name to matter or put their name to a work of art. If an artist possesses a refined sensitivity to matter, then the accumulations of the scrapheap or landfill site are surely a political affront to their own private ecologies of stuff. The word, 'gravity' is not only synonymous with an enigma, but conveys meanings of danger, seriousness, and weighty dignity ('gravitas'). 'Gravity' and 'gravitational' are generally conflated, but gravity is related to the terrestrial pull on weight with the word 'gravitation' reserved in its Newtonian meaning for celestial motion. Gravity is invisible and enigmatic, yet its effects are ubiquitous and obdurately a part of everyday experience. In effect, it is a relatively weak force, allowing certain pleasures to be enjoyed in its trap. Our balloons, swings, trampolines, and ball games create momentary levitations and mock flights. Nevertheless, for returning astronauts, gravity is, in the initial period of readjustment, a powerful force. As they return to earth's gravity, they commonly describe the intense feeling of weight that hangs over them like an iron cage with tongues and lips heavy in the mouth.

As a primary condition of being, gravity is mediated here through art and through the body, rather than filtered through an abstract-scientific discourse. It is perhaps our complete conditioning by gravity and its foundational status within the field of physics that has muted a cultural or aesthetic discourse. Visual perception could be said to eclipse this fundamental and personalized sensation of gravity with sight leading us upward and outward in contrast to the downward and inward of weight. Weight is hard to recover from the deep stratum of the body, its effects not easy to extract in its general diffusion through embodied experience. Certainly, with twentieth-century developments in phenomenology, artists started to incorporate critiques of pure optical perception so that experiences of the body, such as the blind sensation of weight, touch, and movement were reclaimed from sight's dominance.

Our empirical understanding of gravity is experienced through states of falling and acceleration. However, if an inanimate object or thing falls randomly, then it also harbours a premonition of its own agency or will. Conversely, in the act of falling, the human subject slips out of fixed identity and becomes object-like. The fall is normally a chaotic experience where the will is overtaken by gravity. Artist Bas Jan Ader was always keen to stress that his falls were precipitated by the fact that gravity had made itself master over him at that particular moment. However, there are other subtle connotations in the relationship between gravity, falling, and the will. Michel Serres' *The Birth of Physics* (1977) signals a move away from predictable Newtonian mechanics to revalue the voluptuous atomism outlined by Lucretius (99 BC–55 BC) in his extended poem, *On the Nature of Things*. In his account, Lucretius presents creation as dynamically deviant with causality being free from God and springing spontaneously from the law of nature. This is argued through his reading of the *clinamen*. In a preexistent rain of atoms, the smallest deviation from the dead vertical fall produces accidental mixture and variation. This swerve becomes a model of free will, liberated from the chain of causality so as to admit infinite variety and constant change.

Although there has been no comprehensive written account of how gravity and its effects have become significant in contemporary art, there have been a number of curatorial projects with many small exhibitions hosted around the theme in recent years within both commercial and public contexts. My thumbnail references here are by no means exclusive. In an exhibition called, *Dancing on the Ceiling: Art and Zero Gravity* at Curtis R. Priem Experimental Media and Performing Arts Center, New York, in 2010, Kathleen Forde curated a mix of film, dance, photographs, and sculpture, including the work of *Arts Catalyst*, an organization which tenders artistic projects for weightless environments created on parabolic flights. The weightless or dematerializing effects of new technologies and environments served as the focus of an exhibition at Rome's *Palazzo delle Esposizioni* in 2001. *Zero Gravity: Art, Technology and New Spaces of Identity* was a mixed show of thirteen international artists that explored the vertigo, automatism, hypnosis, and destabilization caused by transformations in new media and architectural environments. More recently, in 2011, *Falling Up: The Gravity of Art* was an exhibition mounted at the Courtauld Institute by the MA students, incorporating both historical and contemporary works by artists as diverse as Rubens, Rodin, and Cornelia Parker. In another historical account, artist and filmmaker, Peter Greenaway curated an exhibition of drawings from the collection at the Louvre called *Flying Out of this World* (1994), based on the theme of human-powered flight and falling.

Gravity and Grace: The Changing Condition of Sculpture 1965–1975 at the Hayward Gallery in 1993 is still considered a landmark sculpture exhibition. The show drew together certain threads of post-1960s sculpture, characterized by objects that hugged the floor, leaned against walls, or hung from ceilings by their own weight, all without plinths. *Gravity and Grace* included works by Richard Serra, Giovanni Anselmo, Janis Kounellis, and Barry Flanagan, refracting the sculptural typologies and sympathies of *arte povera* and post-minimalism. Jon Thompson curated the exhibition and, at the time, had been instrumental in developing the MA Fine Art at Goldsmith's that launched the YBA scene. Nearly twenty years later, in 2012, three contemporary female artists (Rachel Howard, Jane Simpson, and Amelia Newton Whitelaw) formulated a response to *Gravity and Grace* and its foregrounding of material and process, in a renamed exhibition, *Gravity and Disgrace* at Blain Southern. Predating Thompson's *Gravity and Grace*, kinetic artist Gerhard von Graevenitz curated an exhibition called *Pier+Ocean* at the Hayward Gallery in 1980 based on three key themes, the final section being devoted to gravity, represented by Carl Andre, Bas Jan Ader, Richard Serra and Sigurdur Gudmundsson amongst others.

Guy Brett's exhibition, *Force Fields: Phases of the Kinetic* held at the Hayward Gallery in 2000 formed an interpretation of art since the 1960s based on an enlarged concern for the forces of gravity, light, magnetism, and electricity with artists playing with larger concepts of space and time and rethinking their practice through a dialectic of matter and energy. In my interview with Brett some years ago, he spoke of the importance of

Alexander Calder's *Circus* constructed in the late 1920s and performed to the early avant-garde in Paris during the early 1930s. Calder had been trained as an engineer but became interested in the circus's gravity-defying performances within a force field, demarcated by the circus tent. His little wire figures of acrobats and trapeze artists became conduits into his abstract sculptures of taut wire and moving shapes. Calder's use of circus is part of a more general appropriation of the circus, both aesthetically and philosophically with its particular disposition to gravity and risk. Circus has subtly informed aspects of art and performance in terms of choreography and movement from the early to mid-twentieth century onwards. Stephanie Rosenthal curated an exhibition entitled, *Move: Choreographing You* at the Hayward Gallery in 2011, which included equipment that allowed the audience to swing, play or hang on suspended ropes. Within the archive selection of nine key themes, around eighteen works are placed under the rubric of *Gravity/Falling* as represented by many of the artists in this book, such as Bas Jan Ader, Trisha Brown, Yves Klein, Gordon Matta-Clark, Bruce Nauman, Steve Paxton and Richard Serra.

Questions emerge as to whether such conscious arbitrations of gravity are just coincidental with other art historical readings. Gravity's invisibility has been mixed with processes of diffusion or evaporation into air as in the dematerialized practices of the late 1960s onwards, emergent in a mix of late romantic irony or 'sublime' conceptualism. Ephemerality has been politically significant since the 1960s, forming a rejection of materiality's authoritarian or commodity aspect. More recently, the idea of precarious practices, as explored by Anna Dezeuze in her book, *Almost Nothing* seems contingent to my study. Dezeuze has argued for a trope of precarious or fragile practices in contemporary art. She argues that the thrown-away quality of the junk aesthetic or process performance constitutes its fragile status as 'almost nothing', again challenging the object-hood of the commodity artwork, whilst avoiding both the vagaries of the sublime or the certainties of form's formality.[3] Yet, in contrast to these interpretations,

---

3   Anna Dezeuze, *Almost Nothing: Observations on Precarious Practices in Contemporary Art* (Manchester: Manchester University Press), 19.

there is more often a 'some-thing' in my examples, an enduring solidity present in the gravity or weight of a thing. Although gravity is invisible, its effects are rooted in a substantial and fundamental experience or value of the world. Gravity costs nothing – it is a sort of democratically available 'readymade'; nevertheless, its action can be completely transformative and deeply political. By unlocking this immaterial material, artists such as Richard Serra, Robert Smithson, Wood and Harrison, Catherine Yass, Bas Jan Ader, Cornelia Parker, and Santiago Sierra enlist gravity as an engine of creativity, play, and risk-taking performance in the manner or look of a circus trick.

How else have artists consciously integrated gravity's downward pull into their creative processes, materials, and performances? I investigate how gravity forces some fascinating intersections with falling that acquire the value of a practice or trope, and how these in turn recoup certain traditions within vernacular entertainment such as circus, vaudeville, and film. In its antithesis to frivolity or lightness, gravity is a serious business, a force to be resisted in the way it is seized in engineering, physics, and architecture, as well as in the way we counter gravity in our orientation to uprightness and the orthogonal environment. By subverting gravity's instantiation in buildings of power and systems of regulation, artists have returned to a play ethic in order to unseat the assumptions and habits evident in these environments and behaviours. From the early modern avant-garde moment, circus has been an intriguing diversion for artists, synonymous with an outlawed or exiled condition. In some way, the body's ability to defy gravity in acrobatics and trapeze is agonistic to values of physical and moral rectitude. Circus's kinetic spectacles segued into early film, comedy burlesque and vaudeville routine and these forms of mechanical entertainment have proved vital sources of inspiration for many of the contemporary artists discussed here. Weight is tied to low cultural forms such as slapstick and the clown's pratfall, described wonderfully by Kreider + O'Leary as a 'frenzy of the material world'.[4] Burlesque comedy of stage and screen proved an inspiration to modern artists and writers as forms of cultural degradation

---

4   Kristen Kreider and James O'Leary, *Falling* (Ventnor: Copy Press Ltd), 51.

that reanimated their work. Heinrich von Kleist provides a slightly earlier augury of this perhaps more modern process of delving into low cultural forms as a way of reinvigorating high art practice. His speculations on the vernacular form of marionette theatre (*On the Puppet Theatre*, 1810) anticipate modern understandings of grace as mediated by anti classical ideals.[5] Although I rely on an easy transition between contemporary art and vernacular entertainment, I still preserve a whiff of the aesthetic, drawn from Theodor Adorno's *Aesthetic Theory*.[6] Adorno draws an equation between art's enigmatic force and the circus *tour de force* or balancing act. Given the earlier fascination for circus shown by artists such as Picasso, Degas, Rodin, Seurat, Toulouse-Lautrec, and Léger, I preserve the word 'modern' and 'modernist' in the way that it is informative of contemporary practice now.

Gravity is everywhere all at once. So how do artists distil gravity in their work? I turn to a musical metaphor to explain my method of selection. For an artist such as Richard Serra (discussed in Chapter 3), gravity has been the principle theme of his work over a lifetime, gaining in significance through a steady of accretion of mass in his sculptures. Gravity is also the foundation of Bas Jan Ader's series of falling works from the early 1970s and was indelibly linked to his particular blend of pathos and metaphysics. For Robert Smithson, the main focus of Chapter 4, gravity is an important leitmotiv or counter theme in his reading of geological time and entropy, whilst for Cornelia Parker, it has been an equally consistent and fundamental theme. In similar concentration, gravity is a fugue voice to Catherine Yass's main fascination for the psychology of space and its impact on human identity. In the work of other artists such as Stan Douglas or Fischli and Weiss, gravity's effects are the trills and grace notes, providing leverage into deeper considerations of magic, agency, matter, and demiurgy. Into this counterpoint, I weave examples that apply gravity as ground bass, second theme, leitmotif, and ornament.

---

5   Heinrich von Kleist, 'On the Puppet Theatre' in *Selected Writings*, ed. and tr. David Constantine (Indianapolis: Hackett Publishing Company Inc., 2004), 411–16.

6   Theodor Adorno, *Aesthetic Theory*, tr. Robert Hullot-Kentor (London: Continuum Impacts, 2004).

Gravity is a dialectical condition, a struggle between heaviness and lightness, resistance and liberation. My chapters are therefore divided into various themes, structured as follows. Chapter 1, 'On Shaky Ground', is about falling and its psychological ramifications within the city environment. I examine how artists such as Catherine Yass, Gordon Matta-Clark, Yves Klein, Jeremy Deller, and Ilya Kabakov, amongst others, have used the body to disrupt urban architecture. Enlisting the antic freedom of film's comedy burlesque or the circus performance, artists and performers subvert the habits and reassurances of sky-high architecture. Oscillating between art and the vernacular traditions of circus, the tightrope walk is examined as a recollection of Nietzsche's 'dangerous wayfaring', somehow suggesting a metaphor for art's own precarious condition.[7]

Chapter 2, 'The Cosmic Cage', records how falling is used as an engine for creativity and play, reflecting on how contemporary artists have manipulated the idea of 'prime mover' or hidden cause. In examples drawn from Stan Douglas, Bruce Nauman, Bas Jan Ader, Fischli and Weiss, and Rodney Graham, the mishap of falling reveals the nightmare of eternal return. The comic double acts of Gilbert and George, Wood and Harrison, like the famous Beckett double acts, are tied together in hellish routines, operating in tandem. By enacting the role of demiurge, the artists in this chapter use gravity as occult force to confound expectations of agency and control. Falling is activated as a cosmic joke or magical trick in ad hoc assemblages of objects and bodies.

In Chapter 3, 'Heavy Stuff', I change direction to focus on the importance of gravity within the world of heavy industry and the technological sublime. As Raymond Williams remarks in his essay, 'Culture is Ordinary': 'At home we were glad of the Industrial Revolution', reflecting a working-class nostalgia for the machines and engines that made work easier and life better.[8] However, the exploitation of the factory worker also produces a collective shame and guilt about the alienating effects of hard, repetitive

---

7   Friedrich Nietzsche, *Thus Spoke Zarathustra: A Book for Everyone and No One*, tr. R. J. Hollingdale (Harmondsworth: Penguin, 2003), 43–4.

8   Raymond Williams, 'Culture is Ordinary' (1958) in Ben Highmore, ed., *The Everyday Life Reader* (London: Routledge, 2002), 97.

labour. This chapter explores how and why contemporary artists have navigated the world of heavy industry and identified with the role of manual labourer, together with how gravity's relationship to mass-as-weight exists in correlation to mass labour and mass culture. My cabal of circus emerges as the leisure-time counterpart to industrialized society, exertion, and effort. Artists such as Richard Serra, Theaster Gates, Cornelia Parker and Chris Burden transform mass into poetry. With tipper trucks, steamrollers and fire engines, their poetic re-inventions of industrial machines dissociate a mechanics of force from the imperatives of capitalist industry, whilst still preserving the historical consciousness associated with work and heavy industry.

In my penultimate chapter, I turn to gravity through the metaphor and condition of vertigo. For Robert Smithson, the twister of the American flatlands, most famously evident in the 1939 film, *The Wizard of Oz*, is a revelation of 'the agony of gravity'; however, for Smithson, vertigo also relates to the spiral mechanics of cinema that peddles mere illusion to produce a hypnotic state akin to dizziness. With his particular use of materials such as glue, asphalt, and rocks, the artist used the provocation of cliff edges and inclines to encourage their collapse. Smithson's invocation of gravity is connected to his interest in entropy, but is also handled through his understanding of Gnostic doubt, particularly evident in his descriptions of Art Deco's mirrored interiors and Hollywood's lurid Technicolor films.

My final chapter, 'Critical Mass', is about the body, sculpture, and performance, taking examples of works by Robert Morris, Yvonne Rainer, Steve Paxton, Allan Kaprow, and Claes Oldenburg. It explores how gravity became overtly expressed within dance and sculpture of the 1960s as an experiment with the body's natural physics. Along with ideas of agency reduction, the non-flourished quality of Judson dance alienated the subjective, authorial, or idealized body, instead emphasizing commonalities between amateur and professional, flesh and object in gravity's democratic conditioning of the body. Subverting subjectivity, intentional agency, and classical ideas of upright carriage and grace, performers and artists have changed key co-ordinates in art and performance, asserting the qualities of exertion, of Sisyphean tests, ordeals, and chance.

In falling for gravity, artists since the early 1960s have enlarged our feeling for weight, mass, and balance, rupturing this template of normal experience to expose the naked contingency of being gravity-bound. Key themes emerge through my synthesis of these projects, ideas and philosophies in the way artists play with gravity, with readings of contemporary art filtered through the popular cultural traditions of circus, vaudeville, and film; and through ideas of atomism, Gnosticism, new materialism, and existential philosophy. With this book, I hope to pull the force of gravity from its cultural hinterland to a more visible and tangible position within the context of contemporary art and performance.

CHAPTER 1

On Shaky Ground

In this chapter, I explore how artists have used gravity, examining a range of performances and works, from tightrope walks to mock flights and other dramas of the lost footing. My context is sky-high architecture and the urban setting. Rising vertically, the skyscraper is a potent counter-gravitational image in which load is transmitted through foundations to the ground. Pillars and caryatids were traditionally part of the architectural syntax of gravity, exposing weight and load; however, with the nineteenth-century introduction of steel frame construction, architecture's language of gravity was sublimated into sheer walls and vast planes without externally articulating a building's load bearing. With the application of new technology and engineering, the daunting skyscrapers of New York altered the body's relationship to height fundamentally, inspiring dreams of flight and nightmares of falling. As the quintessential image of urban modernity, the New York skyscraper induces spine-tingling vertigo and intimations of the suicide's fall. Vertiginous buildings are not just mesmerizing symbols of power, they also provide passage to imaginative and subversive play as revealed in the art and performances that follow.

It is the very meaning of the word, 'skyscraper' that carries particular clout; less the architect responsible for a design that scrapes sky or clouds, but more the artist or performer who disrupts the tall building by daring to scale its heights. If modern skyscrapers sublimate architecture's language of gravity through hidden steel construction, then the contemporary artists discussed in this chapter uncover gravity's vanishing act in architecture by invoking or literalizing the fall as imminent threat. Skyscraper architecture is perhaps one of the most visible extensions of human agency and ego, which is possibly why artists and performers have often been tempted to intervene. Other disruptions flow from my examples in relation to vernacular forms of entertainment, such as popular film and circus, which quietly infiltrate works and performances of contemporary art. As such, I feel no uneasiness in terms of the penetration of popular and sensational forms into the tough constraints of art's ideal or aesthetic form, since art itself provides a constant redefinition of itself through these vernacular practices, dubbed 'the prose of the world' by contemporary art philosopher, Jacques Rancière.[1]

Artists such as Yves Klein, Gordon Matta-Clark, Ilya Kabakov, and Catherine Yass have revealed, rejected or imagined the force of gravity in acts or simulations of flying, falling, and balancing invoked through different conjugations of bodies and buildings. These artists have consciously situated their works and performances within particular architectural settings or contexts so as to disrupt their normal taxonomy of function. In these works and performances, the body iterates or exposes the structures or frames of the buildings, confounding their form or inertness through live action, events and physical movements that illuminate the danger of gravity, as well as its pleasures.[2] The historical range includes Yves Klein's well-chronicled leap in 1960 to Jeremy Deller's most recent project, *Sacrilege*, which coincided with London 2012, but I also delve explicitly into the context of modernity, given my architectural brief and in light of the fact that so many contemporary artists still play a game with modernism.

In forms of art, performance, film and photography, the co-ordinates of sky-high architecture provide the backdrop for a variety of performances that mediate gravity in both comic and serious fashion. Artist Catherine Yass made a film of a death-defying tightrope walk across two tower blocks at a high-rise housing estate in Glasgow in 2008 (*High Wire*). The vertiginous comedy films of Harold Lloyd from the early 1920s also inspired Yass to create a series of works based on the scene from *Safety Last* (1923) in which Lloyd dangles precariously from an architectural clock, his weight dragging time backwards as he clings to the clock hand. Yass's funambulist for *High Wire*, Didier Pasquette was taught by the most celebrated

1   Jacques Rancière, *Aisthesis: Scenes from the Aesthetic Regime of Art*, tr. Zakir Paul (London: Verso, 2013), x.
2   Bernard Tschumi, *Architecture and Disjunction* (Cambridge, MA: The MIT Press, 1996), 3.

contemporary tightrope walker of all, Philippe Petit. Petit's extraordinary Twin Tower traversal in 1974 was more recently explored in James Marsh's documentary film, *Man on Wire* of 2008, and displayed at Mark Wallinger's Hayward exhibition, *The Russian Linesman*, in 2010.

Long before the shadow cast by 9/11, the new skylines of modernity have sharpened our sense or fear of falling. Architectural theorist, Anthony Vidler identifies distinct urban psychological states linked to altitude, listing the following phobias recognized by the end of the nineteenth century: 'cremnophobia' (fear of precipices), 'acrophobia' (fear of elevated places) and 'stasophobia' (fear of vertical stations).[3] According to Vidler, these pathologies were specific to the new urban contexts; however, Gaston Bachelard argues for a universal psychology of gravity in terms of an '*engram* of the immense fall', a template of experience that lurks in the psychological space between body and memory from infancy onwards.[4] In the unequal pairing of frail body and powerful building, a knife-edge exists between laughter and fear. In this David-and-Goliath standoff, the body playfully uncovers the hubris of architecture in what I describe as an architectural burlesque. Tightrope walking is a kind of tease, taunting the leaden-booted pedestrians below, whilst undressing the building's scale and gravitas through physical bravado. Its relationship to circus's tawdry sublime taints the architect's ideal flat plan.

This mix of pathology and psychology might explain the need for a cultural processing of such fears and fantasies, screened onto the New York skyline in film, photography and the arts. Falling became a psychosocial reality, and vertigo, that dizzy preparation for the fall, was rehearsed in the common imagination. The city's early skyscraper architecture with its viewing platforms, rooftop nightclubs, penthouses, and other vertiginous outposts provided elite views for tourists, photographers, and camera-people. A rush of curiosity for sky-high architecture mapped by film and photographic images was also accompanied by fascinations and anxieties about how the vulnerable body could or should respond to such imposing

3   Anthony Vidler, *Warped Space: Art, Architecture and Anxiety in Modern Culture* (Cambridge, MA: The MIT Press, 2000), 35.
4   Bachelard, *Earth and Reveries of Will*, 264.

elevations of architectural mass. Cultural mediators such as King Kong or Superman translated the scale and power of this new terrain and Lewis Hines photographed the aerial cowboys and derricks, responsible for building the Empire State Building in the early 1930s. Buildings themselves are never innocent bystanders of course. In relation to his series *Advertisements for Architecture*, architect and theorist Bernard Tschumi considers how buildings dissemble and disguise their true dispositions:

> Architecture resembles a masked figure. It cannot easily be unveiled. It is always hiding: behind drawstrings, behind words, behind precepts, behind habits, behind technical constraints. Yet it is the very difficulty of uncovering architecture that makes it intensely desirable. This unveiling is part of the pleasure of architecture.[5]

Burlesque performance is characterized by the erotic striptease and thus, for Tschumi, the difficulty in stripping off architecture's clothing is actually formative of its pleasure. So how does Tschumi's masked figure creep into the corners of our experience, shaping or expressing this sense of gravitational force or *force-sense*? How does architecture veil or reveal our most basic instincts for, or sensations of collapse? Gordon Matta-Clark suggests that we think about 'shelter' as a 'form of being' so that falling is contrasted with our most fundamental 'sheltering' instincts: our need for enclosure, stability and uprightness.[6] Perhaps we can say that architecture represses our own trembling condition and the molten condition of the earth we stand on by impressing its language and image into forms of stability.

Gravity is a literal grounding force and also a continual metaphysical re-grounding. It is both obtusely an attraction to the ground and a magical force of strange origin. And yet from such contrary eruptions of meaning, gravity somehow became contracted to a law synonymous with the very notion of firm foundation. From the seventeenth century onwards, the law-giving quality of Newton's theory became the very model of a universal law.

5   Tschumi, *Architecture and Disjunction*, 94.
6   Kirsty Bell, 'Gordon Matta-Clark', *Frieze*, 78 (October 2003) <http://www.frieze. com/issue/review/gordon_matta_clark/>, accessed 8 March 2016.

Gilles Deleuze and Félix Guattari explain how universal attraction became the law of all laws, 'in that it set the rule for the biunivocal correspondence between two bodies; and each time science discovered a new field, it sought to formalize it in the same mode as the field of gravity.'[7] Gravity symbolized the very model of reason, whilst remaining the mysterious force it still is. Stephen Walker notes how Newton's gravitational constant has been inscribed within architectural space, binding bodies to the order of its geometry, holding us in place as subjects of architecture:

> This gravity would also covertly underwrite the relationships between 'us' and Architecture, usually peddled as a subject-object relationship with Architecture as an inert object. However, such a gravity also works to maintain us as the 'subjects' of architecture, subjected to it, while Architecture maintains itself, as much as it might deny it, as subject or discipline.[8]

Gravity produces a template for normative spatial experience and so inscribes habits or routines of behaviour. Drawing on Paul Connerton's ideas of posture, gravity and social memory, Walker reflects on how the body was yet another object to be mechanized according to the natural sciences:

> The particular laws of nature governing the movement of bodies per se were progressively applied to the behaviour and interaction of human bodies. Natural laws quietly governed social norms, where the behaviour of the materialised human body became increasingly predicated on 'the propriety of gravity and the upright viewer.'[9]

Just as gravity is invisibly but indelibly ingrained within buildings, space, and the behaviour of bodies, it is also embedded within philosophy. The powerful conceptual knot that binds philosophy and architecture is tied

7   Gilles Deleuze and Félix Guattari, *A Thousand Plateaus: Capitalism & Schizophrenia*, tr. Brian Massumi (London: The Athlone Press Ltd, 1999), 370.
8   Stephen Walker, 'Baffling Archaeology: The Strange Gravity of Gordon Matta-Clark's Experience-Optics', *Journal of Visual Culture*, 2 (2003) <http://www.sagepublications.com>, 161–85, accessed 18 February 2016.
9   Stephen Walker, *Gordon Matta-Clark: Art, Architecture and the Attack on Modernism* (London: I. B. Tauris, 2009), 94.

to the very notion of ground with a parallel between epistemic ground and literal ground. Philosophy is in essence always founded on an abyss or on shaky ground, because it has a continual and on-going recourse to the grounds or foundation of an argument. Descartes's foundationalism, which drew on an architectural analogy, insisted on the need to insert doubt and questioning into philosophical thinking. He uses the metaphor of bulldozing the ground prior to building a piece of architecture so that rather than the certainty of firm foundations, it is the abyss or doubt that opens up philosophical enquiry.

It is such states of collapse that expose the so-called unshakeable ground for thinking. Binding gravity into scientific discourse as foundational law was crucial to locking institutional, philosophical, and aesthetic foundations into place. Gravity is a trans-locational force that binds bodies together into spaces and buildings along rational assumptions of uprightness and proper behaviour. The power of gravity lies in the way it subtly underwrites spatial-behavioural codes, yet the way it forms such codes is rarely uncovered. Since gravity is fundamentally connected to the language of science, philosophy and rationalism, any subversions or insurrections against its power are highly subversive acts. In some of the examples that follow, artists such as Gordon Matta-Clark and Catherine Yass expose the way gravity is coded in architecture, examining its connections with power.

As skyscraper architecture turned streets into sunken wells, the airy heights became fantasy symbols of freedom and power, as well as fear. Psychological survival meant exceeding the boundaries set by industrialized modernism, testing the extreme discordances of scale and height it had produced. Tall buildings alienate the body, producing a psycho-dramatic imagination of falling or of being crushed by vastness and weight. In thinking about how the body is integrated into the domains of city experience and its architecture, other dreams of universal freedom, play or solitary resistance arise. Let us imagine in the following artworks and performances that gravity is a sort of medium and that architecture provides the stage, antagonist or prompt for real or faked leaps, flights and gravity-defying feats. Falling is antithetical to architecture, a space of experience that deconstructs the tower or skyscraper.

Stepping Off

The work of British artist Catherine Yass examines just such a powerful intersection between height, gravity, psychology and architecture. In *Descent* (2002), part of her Turner Prize exhibition, Yass presented an 8-minute film of a slow disembodied descent down the side of a corporate skyscraper still under construction at Canary Wharf. In contrast to the rapid ascent of capital and the corporate skyscrapers that give it visible form, Yass's film enlists slowness and downward drift to destabilize this urban landscape and its contextual meanings. Opening in a sea of fog, the film takes us down the building past the open grids and framework to the left-hand side of the film's frame. Yass diminishes her own agency in the filmmaking process by using a mechanical crane to move the camera on its journey downwards, rotating every frame 180 degrees to augment the disorientation felt between head and feet. This slow destabilization of ground is apparent also in the accompanying light boxes, in which she cross-processes positive and negative images, creating the effect of warped space. As the camera moves glacially down the side of the building, the ground appears to shift as in a parallax.

Descent was filmed at Canary Wharf at a time of stock-market volatility when disaster and risk collided. Yass questions the mental space of the fall, during which one is alive but dead and identity is eroded. Through the use of slow motion, she examines the moment where we might turn from active subject to falling object.[10] It is hard to envisage the chaotic sensation of falling from height both physically and psychologically, but the height of the drop deepens the void and adds duration to terror. In his book, *Falling*, Garrett Soden confirms the link between fear and height, describing Gustave Eiffel's consternation when his workers demanded extra money for working at extra height, even though death was as certain at 40 metres as at 300 metres:

<hr>

10 Catherine Yass, 'Keynote Speech', *Vertigo in the City: Conversations between the Sciences, Arts and Humanities*, University of Westminster, 29–30 May 2015.

> ... Eiffel was overlooking a psychological effect that occurs whenever a human imag-
> ines a fall: we sense how long it will take. Our 'terrified imagination' plummets more
> than seven times longer from three hundred meters than from forty multiplying
> sevenfold the terror we would expect to see in the air.[11]

Risk and fear challenge architectural assumptions of order, routine and stasis. Bernard Tschumi, taking his lead from Bataille, notes the way that architecture has always held the contradictions of abstraction and experience together. He holds that it is only by doing violence to a building that the real meaning of architecture can emerge: those dramatic personae of death, eroticism and decay allowed to leak through the smiling façades.

In one of his *Advertisements for Architecture* (1976), he uses a film still of a person being pushed from a high window ledge. The subheading for the 'Murder' advertisement commences, 'architecture is defined by the actions it witnesses, as much as by the enclosures of its walls' further framed by the statement, 'to really appreciate architecture, you may even need to commit murder'.[12] Suspicious of architecture's reliance on morphology, function and enclosure, he suggests strategies of dis-programming, trans- or cross-programming architecture with events or bodies moving beyond the limits of buildings, citing certain discontinuous examples such as 'battalions skating on tightropes'.[13] This image suggests discordance between weight and an ultra thin surface, but also reveals how action, body, and event disrupt categories of function and form. Transposing Bataille's reading of transgression and eroticism, Tschumi sees cross- programming as a way of contaminating architecture's rhetoric of purity and form. This he likens to an act of 'cross-dressing'.[14]

Urban adventurers are instrumental in cross-programming or 'cross-dressing' architecture. Ludic interventions such as those staged by high-rise funambulists and climbers disrupt the symbolism and functionalism of the buildings they conquer. In tracing some of the funambulist's faint steps across twentieth-century art and culture, I want to highlight how

11   Garret Soden, *Falling* (New York: W. W. Norton & Co., 2003), 98.
12   Tschumi, *Architecture and Disjunction*, 100.
13   Tschumi, *Architecture and Disjunction*, 186.
14   Tschumi, *Architecture and Disjunction*, 205.

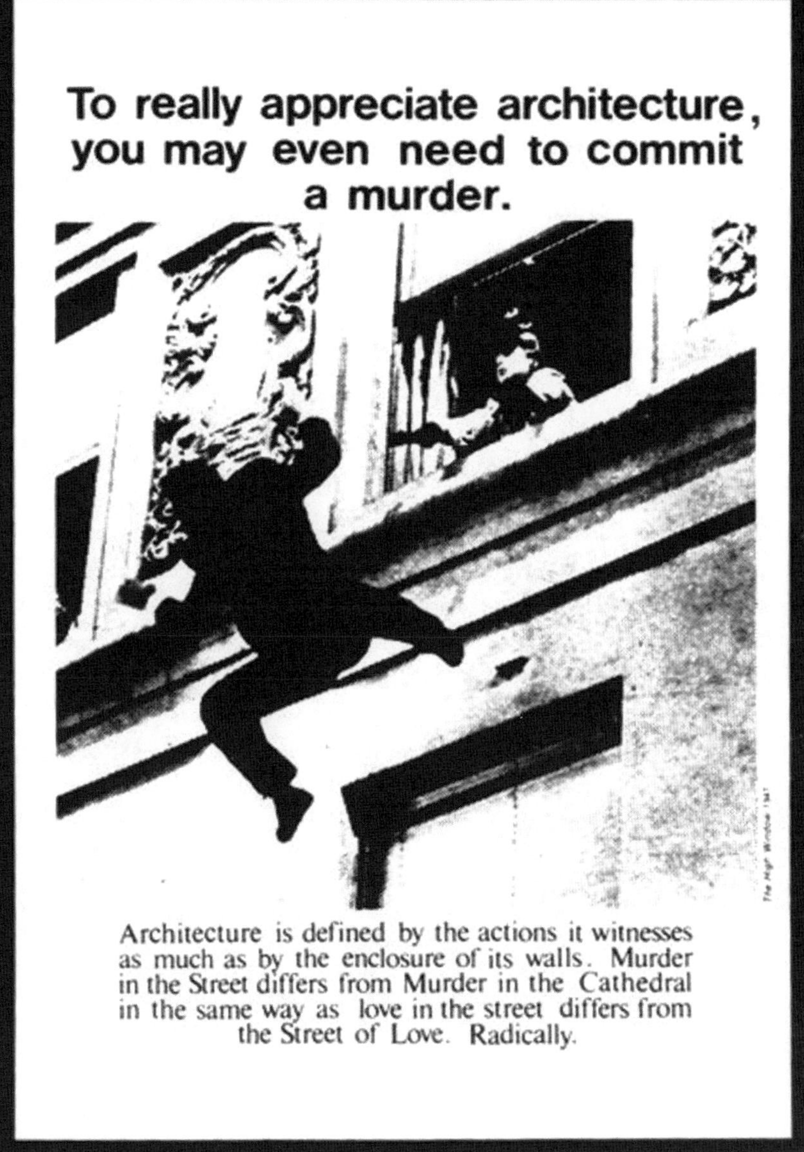

Figure 1. Bernard Tschumi, *Advertisements for Architecture*, 1976.

the body balanced on a wire between two skyscrapers is not just an image of death defiance or libidinal freedom, but is also a momentary revelation of architecture, a sudden but vital striptease that exposes a building's fragility to our momentary perception. This revelation is also evident in film, in the ability of the camera to transcend gravity, a quality extolled by the Russian filmmaker, Dziga Vertov and described in Walter Benjamin's observation of how film's free swooping and gliding exploded the 'prison-house' of urban spaces.[15]

Precariousness unsettles the architect's enterprise, disturbing what Tschumi sees as the 'belief in the unified, centered, and self-generative subject, whose own autonomy is reflected in the formal autonomy of the work.'[16] The sight of a body balanced on its limit or brink opens the building up to the potential of its own collapse (a haunting prolepsis in the case of Petit's tightrope walk). Petit's well-documented walk between the Twin Towers included lying down on the rope for a rest and 'cocking a snook' at a policeman, powerless to arrest him. His impish act undid the certainties of the Twin Towers, breaching their limits by roping two buildings together and extending their boundaries by threading a lifeline across the void space in between. His performance shadowed death against architecture's core logic of safety, shelter, and security, exposing architecture's silent imperatives of containment and control. If tightrope walking has always been an ambiguous spectacle belonging all at once to the domains of sport, circus, fair or carnival, then its interaction with architecture is similarly disruptive with gestures that dress up, cross-dress, or undress a building.

The tightrope has been consciously appropriated in a number of artworks and exhibitions from Expressionist painting to contemporary performance art. In 1971, Robert Morris included a low-level tightrope in his exhibition, *Bodyspacemotionthings*, at the Tate Gallery, along with giant seesaws, slant boards, beams, ramps, and ledges. Morris had developed a line of practice from his earlier plywood sculptures exhibited at Green Gallery

---

15   Walter Benjamin, *The Work of Art in the Age of Mechanical Reproduction* (1936), xiii, tr. Harry Zohn <http://www.marxists.org/reference/subject/philosophy/works/ge/benjamin.htm>, accessed 2 June 2017.

16   Tschumi, *Architecture and Disjunction*, 208.

in 1963 to large-scale props that invited the audience to calibrate their body against gravity's pull, part of Morris's general turn to participation. Steve McQueen's tightrope walker in *Five Easy Pieces* (1995) is filmed from beneath with the rope cutting into the foot, emphasizing the tension and pain associated with the performance. McQueen describes the tightrope walker as 'the perfect image of a combination of vulnerability and strength'.[17] As in many of McQueen's short films, a close web of association exists with silent, black and white films such as Buster Keaton's burlesque comedies or primitive films that simply tracked different forms of movement. At the beginning of *Five Easy Pieces*, the tightrope dips and vanishes into a black screen, likened by the artist to a sense of 'passing through' as in the road movie or its precursor, the phantom film ride, a filmed train ride popular in early film.[18] Passage is more important, just as tightrope walking is less an arrival than 'an indefinite deferral of destination.'[19]

On 22 July 2007, Yass filmed an unsuccessful attempt by Didier Pasquette to walk across a 45-metre tightrope stretched between two 89-metre tower blocks at Red Road in Glasgow. Powerful crosswinds on that day prevented him from crossing the wire between the two buildings and, after reaching the 20-metre mark, he was forced to abandon the walk, shakily reversing his steps back to the platform. Exhibited in 2008 (11 April to 24 May) at the Centre for Contemporary Arts in Glasgow, the resulting film, *High Wire* and four black and white negative light-box images of the Red Road towers form a meditation on the failed utopian projects of 1960's urbanism. Pasquette's precarious walk between the towers supplied the 'air and grace' that brought these social consequences into view. Developed in collaboration with Artangel, Yass's film was produced as an

17   David Curtis, 'A Century of Artists' Film in Britain' (19 May 2003–18 April 2004) <http://www.tate.org.uk/whats-on/tate-britain/exhibition/century-artists-film-britain/century-artists-film-britain-progra-22>, accessed 9 February 2016.

18   Patricia Bickers, 'Let's Get Physical' interview with Steve McQueen, *Art Monthly* *202* (December 1996–January 1997) <http://www.artmonthly.co.uk/magazine/site/article/steve-mcqueen-interviewed-by-patricia-bickers-dec-jan-96-97>, accessed 2 February 2016.

19   Steven Connor, 'Man is a Rope' in *Catherine Yass High Wire* (Artangel, 2008).

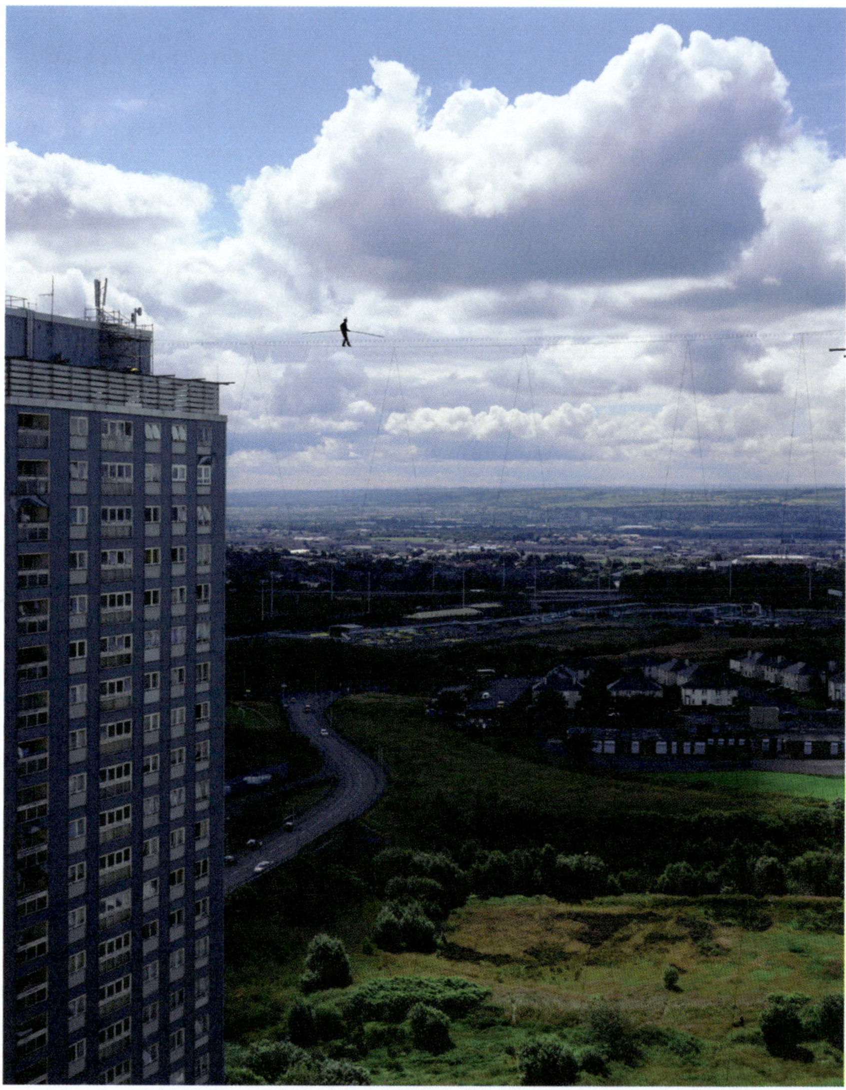

Figure 2. Catherine Yass, *High Wire*, 2008.

installation with four different viewpoints intended to subvert the stable,
fixed viewpoint through our own giddy perception of his tentative walk,
mediated through a camera on the high-wire walker's helmet. Artangel

director, James Lingwood explains: 'Red Road was the perfect place for the conjunction of two different dreams – the architects' and planners' dream of building into the sky, and the individual one of walking into the air: planning, containment and control countered by an expression of space and freedom.'[20] Two distinct ideas are contained in Yass's *High Wire*: the psychology of the fall and the relationship of human identity to the architectural spaces of modernism. I now discuss these themes within a larger framework: the question of how a circus act like the tightrope walk forms a burlesque (comic and erotic) of sky-high architecture.

Collapse

Architecture, with its facades, window-eyes and skin, is anthropomorphic. It resonates with our own sense of embodiment like a second skin wrapped around us. Within the context of post-war Brutalism and its watered down derivations, architecture has been read through the body as a figure of pain: naked and possessing a surface that sweats or excretes damp. These architectural bodies of pain are witnesses to violence, earmarked for assisted death, their social dysfunction seen as unassailable. Brutalism declares its asceticism, the building read as naked body, revealed without mask and lacking any overtures of erotic mystery.[21] Although the term was connected to the French term for raw concrete, Reyner Banham turned it into an 'ism', 'brutalism', which started to convey other meanings connected to a social history of poor building maintenance, poverty and marginalization. Swept into a pejorative reading, brutality was erroneously read into mass housing schemes and generic systems-built tower blocks, paving their way to destruction and demolition at an extraordinary rate.

Completed in 1969, Sam Bunton's Red Road Estate was then the highest residential systems-built mass housing complex in Europe, providing

---

20   James Lingwood in *Catherine Yass High Wire*.
21   Owen Hatherley, *Militant Modernism* (Hants: O Books, 2008), 70.

4,700 new homes for Glasgow's slum-dwellers as part of the city's postwar construction programme. Bunton's warning that residents should not expect 'airs and graces' augured the slow decline of Red Road. However, in *High Wire*, Yass tests the architect's injunction by probing the mixed role of gravity and grace; empirically the life lived on a high-rise council estate long associated with crime, squalor and disenfranchisement and, poetically, the potential for a kind of magic woven out of the sky. She admits being simultaneously repelled by, and attracted to the abyss-fall. Real fires, falls, and suicides occurring over the years at Red Road only added to the decay and deprivation associated with the tower block complex. As Francis McKee explains: 'Red Road, having represented so much utopian yearning now became emblematic of the descent into squalor and despair across all of these schemes.'[22] Far from producing airy aspiration, the social problems associated with estates such as Red Road have been the undoing of much post-war architecture in this country. Mass housing has become synonymous with tumbling tower blocks and, with over 50 per cent of Britain's high-rise blocks based in and around Glasgow, many, including Red Road, have been scheduled for demolition. As the first slab block was detonated in June 2012, a crowd gathered to witness the final act of another utopian experiment turned sour, fascinated by the sublime *telos* of its fall. Yass's *High Wire* shows the physical performance as metaphor and the performer as direct agent, who mediates the possibility of airy grace within a bleak, post-war housing scheme. For the artist, his aborted walk was also emblematic of the position of the residents at Red Road, temporarily and tentatively occupying the air.

Stretched between the Red Road Estate blocks, the tightrope in *High Wire* is the thin, trembling thread linking Pasquette to life. It is also an abstract line scored by Yass across the photograph, opening up the limits of the photographic medium through her literal inscription of the surface by hand. This line in the sky gleams within its dour landscape and together with the manipulation of negative and positive images interrupts photography's spell. Without this physical intervention by Yass, the light box works look more like spectral palimpsests of the architect's design back in the 1960s,

22   Francis McKee, 'I saw a city in the clouds' in *Catherine Yass High Wire*.

stripped of detail, colour, and incident. The funambulist's aborted performance symbolizes the failed integrations of body and space attempted by urban planners. By transferring a circus act to a housing estate, she kindles a playful dreaming about space to challenge architecture's core functions of stability and shelter. The work raises questions of how buildings influence the movement and behaviour of those living there. By stepping out onto the wire, the tightrope walker assumes an outlaw condition, transgressing the grounded-ness that is coextensive with fixed identities and certainties.

The destruction of a Glasgow high-rise complex inspired the Berlin-based artist Cyprien Gaillard to create a number of works, including a 4-metre cenotaph cast from 15 tons of concrete salvaged from the demolition of 12 Riverford Road in Glasgow (*12 Riverford Road, Pollockshaws, Glasgow, 2008*). Displayed at the Hayward Gallery project space in 2014, the obelisk was situated within the fortress-like complex at the South Bank. Within a five-year period, he personally witnessed over thirty demolitions, drawn to the spectacle of falling towers, whilst wanting to recover a rough poetry from the language of modern architecture. Gaillard is fascinated by the violence enacted on modern buildings and the devaluation of brutalist buildings in particular, but his interest also emerges from a structure of feeling, the sort of melancholia that attaches to buildings such as those included in his electronic video opera, *Desniansky Raion*, created with the musician Koudlam in 2007. Gaillard's 30-minute video is composed of three sequences shot in different locations at a number of post-war housing estates, including in the third film, the huge estate of Desniansky Raion just outside Kiev. The final sequence is composed of aerial shots, which reveal the complex's vast scale and sublime aspect. Gaillard likens Desniansky Raion to a kind of modernist Stonehenge, a sort of pathetic fallacy for modernism's failure. Walter Benjamin is indicated here in his allegorical reading of the ruin within city-space, an allegory that makes the dialectic of history visible. Gravity is the quiet engine in this dialectic, a conspirator in the forces of entropy and decline, urged on by historical change.

The demise of post-war housing complexes has also been a subject for artist Rachel Whiteread. Her project, *Demolished* (1993–1996) involved photographing the demolition of generic residential tower blocks on the Clapton Park and Trowbridge estates. Her interest in these sites is tied to a humanist agenda whereby bodies and lives are inscribed into shared communal space. As Tschumi explains, architecture is defined as much by those who witness or inhabit it, as the walls that structure it. Just as these photographic prints represent the skin shed by the tower blocks' image prior to and during demolition, the artist's characteristic use of casting in plaster, resin, and rubber in other projects reifies concealed corners or negative spaces within domestic architecture: voids under stairs, bookcases, bath tubs, or whole interiors are made present in order to reveal the relationship between body and building, personal history and the architecture that contains it. As Victoria Rimell writes of Whiteread's negative casts: 'It is not so much like viewing the inside of a body, as of viewing a skinned body, a stone Marsyas, simultaneously surface and depth.'[23] Whiteread considers the analogies between body and building in her whole negative cast, *House* (1993), which she likens to embalming a body: 'It was like exploring the inside of a body, removing its vital organs. I'd made floor pieces before in the studio, and have always seen them as being like the intestines of a house, the hidden spaces that have generally been inaccessible.'[24] Whiteread's involution of space contrasts with Tschumi's reading of architecture as surface mask or disguise in a programme of repression, the mask being part of a building's ploy of seduction: 'Architecture ... constantly plays the seducer. Its disguises are numerous: facades, arcades, squares, even architectural concepts become the artifacts of seduction.'[25] Spatial and bodily metaphors are ways of disrupting the abstract tendencies of architectural discourse. Tschumi uses the body and its metaphors to invade and disrupt architecture's apparent neutrality.

Falling from height is the preserve of suicides and the most potent transgression of a building's function. As Gareth Soden writes: 'Although death by falling represents a small fraction of all suicides, it is an act that grips our emotions with a power out of proportion to the numbers, because falling carries a peculiar poignancy.'[26] It was Marilyn Monroe's death by overdose that inspired Andy Warhol's *Death and Disaster* series (1962–1967) for

23   Victoria Rimell, *The Closure of Space in Roman Poetics: Empire's Inward Turn* (Cambridge: Cambridge University Press, 2015), 152.
24   Charlotte Mullins, *Rachel Whiteread* (London: Tate Publishing, 2004), 52.
25   Tschumi, *Architecture and Disjunction*, 90.
26   Soden, *Falling*, 102.

which he chose to use a publicity shot from Monroe's 1958 film, *Niagara* in silkscreen prints of the actress. Christian Hite sees a relationship in Warhol's choice of this image and Niagara, a famous suicide spot, as well as a site for famous funambulists such as Blondin. Monroe's death prompted his use of multiplied images, as if he were seeking to empty out the supposed 'interiority' of her subjectivity.[27] This is also borne out in Warhol's silkscreen print (*Fallen Body: Suicide*, 1962) taken from Robert Wiles's famous photograph of Evelyn McHale moments after her leap from the Empire State Building in 1947, dubbed 'the most beautiful suicide' in *Life* magazine. There is a strange anomaly in the suicide's leap: the most famous beauty spots and architectural landmarks are selected as if the deepest introspection demands the most public of platforms. Although Foucault frames suicide as a form of resistance to the technologies of power, there is also an almost bathetic quality to the way the most personal and highly subjective acts leads to the most impersonal of outcomes: 'a tiny bit of brain lying on the sidewalk for the dogs to come and sniff at.'[28] McHale's beautiful face and body hid her insides, which had been liquefied in her dreadful move from falling subject to fallen object. This conversion from subject to anonymous or impersonal matter is refracted through the materials and process of American painting at this time as Warhol's mechanical production line of silkscreen prints (paint pressed into the canvas) replaced Jackson Pollock's drip dances that wove paint, gravity, and heroic subjectivity together.

Architectural Burlesque

Let us return to my key conceit, the idea of the architectural burlesque, fed a transition by Tschumi's concept of cross programming in terms of architecture and the body. How can the notion of 'burlesque' be written into the

27   Christian Hite, 'The Art of Suicide: Notes on Foucault and Warhol', *October*, 153 (Cambridge, MA: October Magazine and MIT, 2015), 65–95.
28   James Miller, *The Passion of Michel Foucault* (Cambridge, Mass: Harvard University Press, 2000), 55.

relationship between performer and building? The word 'burlesque' derives from the Italian word *burla*, a mockery, parody or joke, developing from its seventeenth-century meaning to refer to nineteenth-century variety performances that included slapstick performances and sleazy stripteases. Later in the history of burlesque, the saucy striptease and erotic dance took over, as the slapstick routines of vaudeville migrated into the early film comedies of Charlie Chaplin, Buster Keaton, and Harold Lloyd. Wrapping these variants of meaning around architecture, we might argue that burlesque performance undresses architecture, lampooning its 'gravity' in a sort of deconstructive striptease. The burlesque always concerns a deep discrepancy between the gravitas of its source material and the comic irreverence of parody. In the domain of the skyscraper, fear and parody make intimate bedfellows, if only because we need to make light of our terror of falling.

I invoke this idea of an architectural burlesque through spectacles and performances of the high ledge with entablatures, decorative masonry and clock towers forming the backdrop. A building's frame and ornamental vocabulary are both the limits of the structure and the self-protective buffer zone of its form. Poised on the edge or brink of a high building, the body exposes the architectural frame or limit, making a travesty of its solid presence. In Tschumi's terms of reference, the passage or motion across a building is already a transgression of ideal form and is an act of violence. Playing out the drama of falling in a skyscraper landscape, performers or daredevils subvert a building's function and autonomy. Nowhere is this more evident than in the burlesque comedies of Harold Lloyd, where the 'human fly' dangles and sways on flagpoles, clocks, and girders to make a playground in the sky.

Catherine Yass's more recent film and etchings are based on a scene in the 1923 film, *Safety Last*, when Lloyd, the 'human fly' climbs up the facade of a department store and swings perilously from a clock, dragging the clock's hand back in time. In her film *Safety Last* (2011), she montages nine 12-second repetitions of this short cinematic sequence, adding scratches to the film with each repetition until the image is almost obliterated, leaving only an abstract, striated surface. Just as Lloyd destroys the image of time in this scene, so, Yass suggests, celluloid ages and accrues scratches as it falls repeatedly through a projector.

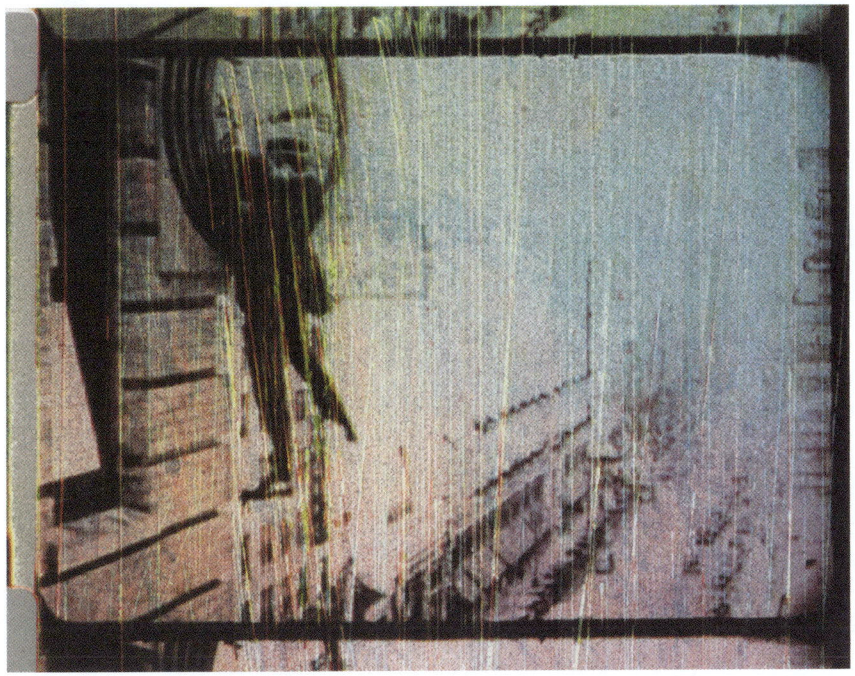

Figure 3. Catherine Yass, *Safety Last*, 2011.

In a half homage to Harold Lloyd's film *Safety Last*, Gordon Matta-Clark's short silent film performance, *Clockshower* (1973) involves the artist donning a raincoat together with white gloves and climbing up onto the clock face of New York's Clocktower. There he shaves, showers and brushes his teeth, precariously harnessed to the hands of the clock. At one point in the film, he clings on to the clock hand, dragging it backwards like Harold Lloyd in the original film. By suspending his body over the clock, Matta-Clark unwinds both its structural presence and regulatory function within city-space. Architectural clocks are visible reminders of the importance of 'clock time' within the city, of the need to co-ordinate the urban work force. Matta-Clark's morning routine orders the body, civilizing it in readiness for its public facing activities, subverted at the same time by risk and a defiance of gravity within civic space. In an irony obviously not lost on Matta-Clark, the Clocktower building was originally built for

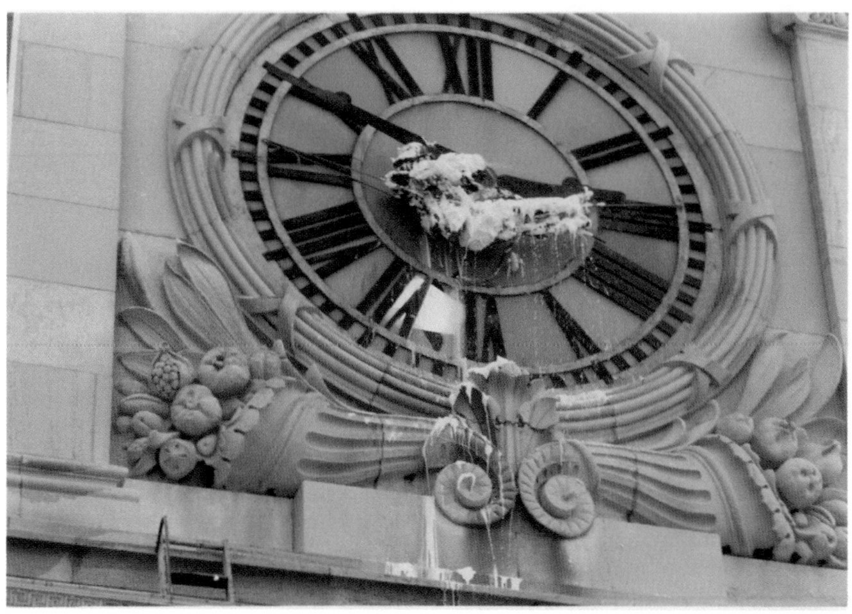

Figure 4. Gordon Matta-Clark, *Clockshower*, 1973.

Metropolitan Life Insurance, a building whose main bureaucratic function was to diminish risk.

Attached to the Clocktower above Broadway, Matta-Clark performed this domestic action in full view of Manhattan. As Stephen Walker writes, 'the film of *Clockshower* nods to a Buster Keaton burlesque' and the schizophrenic mix of public and private in his use of 'burlesque domesticity' tears through the social fabric.[29] Matta-Clark's *Clockshower* performance also reflects Tschumi's disjunctive approach of cross-programming architecture by cutting across the functional grain of a corporate building as he showers and shaves in public view. He emphasizes process in undressing the building's normative function and lacing it with comic absurdity. As Michael Kimmelman writes in his review of the Matta-Clark retrospective at the Whitney Museum of American Art:

29   Walker, *Gordon Matta-Clark*, 117.

The camera pans at the end to show the panorama from Broadway, where Matta-Clark is a speck high on the skyline, nearly invisible. He looks so blissful up there in the soot and sunshine, having what was clearly a great time, reminding us of that less regulated New York City wherein a cheeky young man, if so inclined, might dangle from the outside of a tall building for a while …[30]

Before his untimely death in 1978, Gordon Matta-Clark's radical projects possessed an absurd playfulness in his swashbuckling 'Errol Flynn' approach to deconstructing architecture. Body harnesses and ropes, used also by his friend Trisha Brown in her gravity-defying choreographies, *Man Walking Down the Side of a Building* (1970) and *Walking on the Wall* (1971), were instrumental in his audacious building cuts and slices. In his best-known work, *Splitting* (1974), Matta-Clark cut an abandoned house completely in two, rocking one 15-ton half back on its foundations and removing all four top corner eaves. Of the transformed house, he said:

The realization of motion in a static structure was exhilarating … the suspension and the suspense of it. Putting your shoulder to part of a building, pushing it and having it give way reminds me of silent film comedies, which is the kind of humor I like. Something as primitive as pushing stones.[31]

Matta-Clark deals in edges: the interstitial gaps between an upstairs' floor and the ceiling below, voids within walls, hidden spaces and passageways. By wielding a chainsaw at walls and floors, he cuts intersections into a building so that its precarious contingency becomes clear, with vertiginous drops opening up. As a building became severed or heavily punctured, he would be forced to 'walk the tightrope' across thin gangplanks, beams or fragile staircases to complete the process of deconstruction. He was keen to stress that his athletic movements across a building were a critical element in the process, a dance that activated the building in a continual metamorphosis. His choice of a simple suburban frame house in New Jersey with white

30 Michael Kimmelman, 'Cross Sections of Yesterday: Gordon Matta-Clark Retrospective', *New York Times* (23 February 2007) <http://www.nytimes.com/2007/02/23/arts/design/23matt.html>, accessed 2 February 2017.
31 Gordon Matta-Clark in Corinne Diserens, ed., *Gordon Matta-Clark* (London: Phaidon Press Limited, 2003), 77.

clapboard exterior is the very image of American domestic architecture. Buster Keaton parodied modern pre-fab housing in *One Week* (1920), when he tries, but fails to assemble a flat pack house for his new wife. Keaton's film had in turn been inspired by a short documentary, *Home Made* about prefabricated housing made by Ford Motor Company in 1919. In 2013, Simon Faithfull and Ben Roberts curated an exhibition that reflected the connections between Buster Keaton's burlesque performances and contemporary art, juxtaposing Keaton's flat-pack house disaster, *One Week* (1920) and Matta-Clark's *Splitting*.[32] These performances assault the stasis of architecture with novelty acts and grotesque parodies that draw buildings into a creative dialogue. The human body's tendency to drop or collapse is well matched to lampooning the rectitude of architecture. However, there is another variant of gravity in relation to urban modernity in which the dream of air and space takes on a sublimating role.

Wings of Pathos

Utopian space dabblers are never content to stop at the apex of a tower – they want lift-off. As French writer, Paul Morand wrote after seeing Manhattan in 1925: 'Without roofs, crowned by terraces, it seems to be waiting for airships, helicopters, men of the future with wings.'[33] However, the aspiration for flight is generally a more crudely mechanical, even pataphysical, battle against gravity. The myth of Icarus exemplifies the dream of self-powered flight, providing inspiration for artists from Bruegel the Elder to Peter Greenaway in his 1997 exhibition, *Flying Over Water*, held in Barcelona and Malmö. Icarus symbolizes the dream of flight and the pain of the fall by flying too close to the sun. In *Visions of Excess*, Georges

---

32   An exhibition curated by Simon Faithfull and Ben Roberts, *The World Turned Upside Down* (Warwick Arts Centre, 4 October–14 December 2013).

33   James Sanders, *Celluloid Skyline: New York and the Movies* (New York: Alfred A. Knopf, 2003), 108.

Bataille uncovers the confusion of two psychological conditions connected to the sun in terms of Mithraic ritual: the myth of Prometheus and the myth of Icarus. Icarus flies up to the sun on his homemade wings and then falls to earth as the sun melts the wax:

> All this leads one to say that the summit of elevation is in practice confused with a sudden fall of unheard-of violence. The myth of Icarus is particularly expressive from this point of view: it clearly splits the sun in two – the one that was shining at the moment of Icarus's elevation, and the one that melted the wax, causing failure and a screaming fall when Icarus gets too close.[34]

Spiritual elevation and the fall are part of the same continuum. The dream of ecstatic flight out of bleak importunities on earth has informed the work of the artist Ilya Kabakov. Kabakov's installation, *The Man Who Flew into Space from his Apartment* (1988), part of his series, *Ten Characters*, is a fictitious document of one man's escape from earth into space, using a makeshift sling with springs possibly taken from his narrow folding bed. Comprising a small, down-at-heel Soviet apartment room with the walls lined with images of the Red Square, propaganda, and sketch designs for his crude launch technology, Kabakov creates a *mise en scène* of desperate escape. From the evidence of this narrow room with a hole punctured in the ceiling and the recently vacated sling below, it is unclear whether the man flew into space or, Icarus-like, flew and then fell to earth. In contrast to cosmonaut-heroes such as Gagarin and Viktorov, this imaginary traveller takes solitary flight out of his room and out of the communal space of a failed utopia. In this installation, it is the quality of homespun, DIY technology that parodies the hubris of flight and space technology. Contemporary artists such as Kabakov and Simon Faithfull have built apparatuses that propel bodies or objects into space in mock-tragic anticipation of their eventual rebound back to earth. This trope of the amateur aviator creating homemade aircraft deflates hubris and those statements of

---

34   Georges Bataille, *Visions of Excess: Selected Writings 1927–1939*, ed. Allan Stoekl, tr. Stoekl with Carl L. Lovitt and Donald M. Leslie Jr (Minneapolis: University of Minnesota Press, 2004), 58.

power embedded in hard technology. British artist Simon Faithfull uses low-grade found materials to create his *Escape Vehicles*, navigating the comic pathos of man's fantasy of flight.

From 1996 onwards, Faithfull has mined a peculiarly British tradition of heroic failures in making a series of what he calls *Escape Vehicles*. *Escape Vehicle 2* is a dainty chair made out of old matchsticks with three dead flies tethered above as if ready to lift off, its design inspired by a Victorian plan for a flying machine. The vehicle's delicate structure, described by the artist as 'a truly pathetic object', mocks the propulsive spectacles of real rocket launches.[35] *Escape Vehicle 5* is a 1996 project based at the Royal College of Art, in which a large weather balloon floats in the sky with a chair tied beneath it.[36] The rope is pulled through the skylight of the building, tethering the rudimentary aircraft outside, whilst inside, a pair of scissors is placed behind emergency glass, teasing the visitor into releasing the balloon-chair contraption. Subtitled, *The Campaign Against Living Miserably*, this work invites comparison with Kabakov's installation, both united through a rejection of how gravity ties us to earth. Faithfull's *Escape Vehicle 6* is a 25-minute film of an office chair tied to another weather balloon, launched 18 miles into space. Featured in the *Gravity Sucks* exhibition at the BFI in 2012, *Escape Vehicle 6* is an ecstatic ride into the stratosphere, at which point the atmospheric pressure breaks one of the legs off and the balloon bursts. In an interview with Helen Brown, Faithfull explains how, as a child, he felt melancholy about being gravity-bound and sees his escape vehicles as poignant reminders of the futile hopes of escaping earth.[37] These pathetic contraptions made by Faithfull and Kabakov are deliberately makeshift, the results of imposter hobbyists rather than the technologists of progress.

35   Simon Faithfull in Helen Brown, 'Simon Faithfull's "Gravity Sucks": Furniture's Giant Leap', *The Telegraph* (10 July 2009) <http://www.telegraph.co.uk/culture/art/art-features/5777571/Simon-Faithfulls-Gravity-Sucks-furnitures-giant-leap.html>, accessed 10 August 2012.

36   Simon Faithfull <http://www.simonfaithfull.org/works/escape-vehicle-no-5/>, accessed 10 August 2012.

37   Brown, 'Furniture's Giant Leap', *The Telegraph* (10 July 2009).

In any account of how artists have overcome or been overcome by the force of gravity, it is difficult to ignore the performances of Yves Klein. His showman trick, *Leap into the Void* (1960) combined the mumbo jumbo of mail order Rosicrucianism, Judo, Zen Buddhism, and mood-ideas of cosmic unity. Diving off the ledge of a Parisian rooftop, Klein's ecstatic leap was duplicated in a four-page fake edition of the Paris newspaper, *Dimanche* with the headlines, *Theatre of the Void* and *Man in Space* (1960). His profound ties to the writings of the Rosicrucian, Max Heindel, led Klein into a more secular space-dabbling utopianism, which in the context of the early 1960s space race, had been part of a developing rhetoric within certain artistic groups, such as *Zero*. Famously, Klein proposes: 'We will all become aerial men, we will know the upward force of attraction toward space, toward nothing and everything at once: earthly gravity having been overcome, we will literally levitate in a total physical and spiritual freedom.'[38]

When it comes to artistic expressions of gravity, Klein's leap is a great and much imitated one-liner. He does not overcome gravity but rather plays with the representation of how gravity is conquered, whilst activating space and air in a sort of residual nod to the sublime. In contrast to Bas Jan Ader's falling works of the early 1970s, which take on the very metaphysical injunctions of the fall, Klein's fall is safe in its heroic assumptions of 'hovering in a sovereign Cartesian situation of (masculine) transcendence'.[39] This difference is also evident in the way that Ader *actually* fell and Klein faked his fall. Klein's balancing act between serious artist and playful trickster has lent itself to continual parody. As a student, Paul McCarthy re-staged Klein's *Leap into the Void*, jumping out of a window but falling instead, twisting his ankle in the process (*Sudden Leap*, 1967). In 1971, Harry Shunk and János Kender, who had been hired by Klein to photograph his fake leap, documented Matta-Clark's self-styled response to the leap. Suspended upside down from a beam in a disused building on Pier 18, Matta-Clark's body is made to look like the subject of criminal revenge rather than the

---

38   Thomas McEvilley, 'Yves Klein and Rosicrucianism' in *Yves Klein 1928–1962: A Retrospective* (Houston, TX: Rice University Institute for the Arts, 1982), 241.

39   Andrea D. Fitzpatrick, 'The Movement of Vulnerability: Images of Falling and September 11', *Art Journal*, 66 (4) (2007), 84–102 (94).

embodiment of heroic will. More recently, Ciprien Mureşan photographed his body prone on a pavement in *Leap After Three Seconds* (2004) and, in 2010, Yasumasa Morimura recreated the leap in *A Requiem: Self-Portrait as Yves Klein*. Such parodies and homages inhabit the pregnant vacancy or hiatus of Klein's original leap.

Radical Bouncing

By extolling the air as ideal space or element, Klein rejects the conventional structures of architecture that are too much tied to place and identity, instead postulating an architecture composed of compressed and buoyant air. This is precisely the wrong articulation of air's liberating potential in the terms envisaged by philosopher, Luce Irigaray. In her wrangle with Heidegger, Irigaray argues that the air possesses a radical exteriority and vagrancy, which she characterizes as female. Architecture has always been antagonistic to the air, hardening it into structures of place and rectitude.

Klein's saw the potential of air to become an overarching form of shelter. After his experiments with the void, he turned his ideas and beliefs towards designing life lived in the air, drawing on its buoyancy in an aspiration towards Edenic weightlessness. In collaboration with the architect, Werner Ruhnau, Klein developed plans for a pneumatic paradise, *Project for an Aerial Architecture* (1960), with walls and ceiling replaced by jets of air and wind. He was not alone in proposing a republic built in and on air. Other design groups such as *Utopie, Archigram* and *Haus Rucker Co.* created pneumatic environments and buildings during the 1960s as radical statements of opposition to the managerial rationalities of the modern city. This call to air with its implicit resistance to gravity was rebellious and, again, disruptive of the city's work function.

Artist-engineer Tómas Saraceno refers back to this history and radical moment. Harnessing air for its lightness, he makes installations reactive to gravity, modelled on the fragile power of spider webs, soap bubbles and galaxies, referencing also the geodesic dome of Buckminster Fuller. His practice includes designing and constructing habitable terrains suspended in air, drawing on the sensorial effects of floating in bubbles and on air in order to consider alternative forms of co-living and communication. Immersive environments like *Cloud Cities* (2011), or his plans to make a flying city, are attempts to subvert the functional rationale of the city with its routines of work and administration. In June 2013, Saraceno's vast installation, *In Orbit*, opened at K21 in Dusseldorf with wire webs and transparent spheres constructed in the glass cupola of the museum and suspended 24 metres above the ground. *In Orbit's* light and diaphanous quality means that it is almost invisible unless populated by visitors. As one person moves in a corner of the work, this creates knock-on ripples elsewhere in the giant installation, creating an interesting physical co-dependency. Moving across its surface also invokes the sensation of being held over an abyss, but also helps to weave a connection between those up high and visitors on the ground.[40]

Associated with the inflatable ethos of the 1960s, bubbles and domes have produced other sensibilities and transcriptions of the ideal. *Utopie's* pneumatic environments were partly inspired by Archigram's bubble cities and Buckminster's geodesic domes. The lightweight flexibility of the dome was regarded as a physical metaphor for how to live with more pliant sociability and playfulness, rejecting the obdurate weightiness of *Dasein*. Of the earlier design groups, *Utopie*, formed in 1967, produced one of the most sustained attempts to create pneumatic bubble architecture. Inflatable domes, homes, fashion and furniture were designed for their ethico-political value. Architects, Jean Aubert, Jean-Paul Jungmann and Antoine Stinco, together with left-wing sociologists such as Jean Baudrillard wanted to restore active play to urban centres, which in the post-war period had been reduced architecturally to a drab formalism. Describing their approach to design as a 'joyous critique of gravity', *Utopie* staged an exhibition, *Structures Gonflables* at the *Musee d'Art Moderne de la Ville de Paris* in March 1968.[41]

40   KunstsammlungNWR, 'Saraceno, Tomás-in orbit', [video] (uploaded 21 June 2013) <http://www.youtube.com/watch?v=ROqL-8h_7DM>, accessed 28 February 2016.

41   Antoine Stinco in Marc Dessauce, ed., *The Inflatable Moment: Pneumatics and Protest in '68* (New York: Princeton Architectural Press), 71.

Visionary designer, Buckminster Fuller, the Archigram group in Britain and urban philosopher, Henri Lefebvre were the key influences behind this cathartic re-enchantment of urbanism. Mobility, ephemerality and flexibility were the keynotes of their inflatable landscapes. Bubbles and domes were secondary expressions of levitation, creating micro-gravitational environments to repress the ground. Fuller's geodesic domes were conceived as structures that might combat 'dwelling inertia'.[42] These were structures that ran counter to the dominant international style.[43] Similarly, the *Utopie* group rejected international modernism as having produced so many weary slabs and blocks of uniform housing on the outskirts of Paris.

More recently, British artist Jeremy Deller has discovered the elastic potential of air in his life-size bouncy castle replica of Stonehenge, *Sacrilege*. This blow-up version of the ancient stones recalls the subversive politics of 1960s inflatable architecture. *Sacrilege* was created for the Glasgow International Festival of Visual Art in 2012, timed to coincide with the London Olympics. This global festival of sport makes spectacle out of the battle between body and gravity: pole vaulting, high jumping, long jumping and games of throwing, but in *Sacrilege*, Deller returns simple fun to amateur participants. We return to the context of Glasgow with Deller's bouncy castle located in the former slum area of the Gorbals, sited against a backdrop of high-rise residential buildings. Deller's choice of Stonehenge was informed by a controversial policy introduced by English Heritage of forbidding access to the real site. By contrast, the bouncy castle is open to everyone to climb on and enjoy.

Sacrilege's wobbly ground transforms the city space into an arena for communal play and participation, echoing Robert Morris's invitation to audience participation in his mock playground of *Bodyspacemotionthings*. Just staying upright and connecting to the momentum of its shaky ground is difficult, but the loss of co-ordination produces the greatest pleasure. Bouncing creates a feeling of embodiment that is 'in the moment', allowing us to forget our weight momentarily. Previous to Deller's inflatable Stonehenge, Dana Casperson, Joel Ryan and choreographer, William

42   James Meller, ed., *The Buckminster Fuller Reader* (London: Penguin, 1972), 78.
43   Meller, *The Buckminster Fuller Reader*, 64.

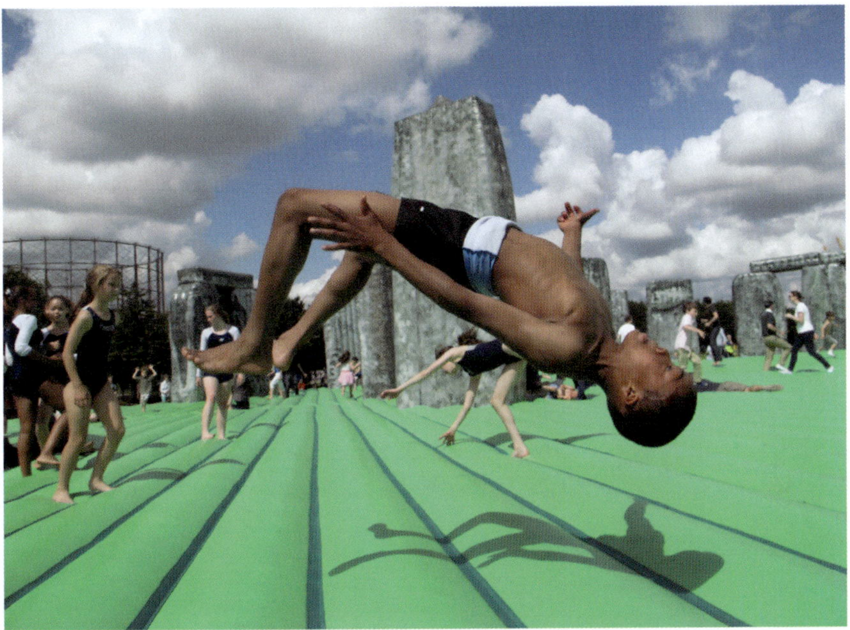

Figure 5. Jeremy Deller, *Sacrilege*, 2012.

Forsythe created a giant bouncy castle installation with Artangel in 1997. *Tight Roaring Circle* was a 40-feet high white pneumatic castle blown up inside the Camden Roundhouse, an old Victorian locomotive shed, complete with floppy turrets and crenellations.[44] Printed in white on the white rubber walls was a poem about failure, impossible to read when balancing over the rippling ground. Like Saraceno's participatory landscapes, bouncy castles create physical co-dependency between jumpers using the same elastic surface. *Tight Roaring Circle* was intended to re-engage the audience as participants in a simple activity that demanded their whole consciousness. Bouncing returns the body to the primal pleasures of simple body-environment interaction, suspending the force of the gravity briefly. As French philosopher, Michel Serres claims, bouncing brings

44   'Tight Roaring Circle' <www.artangel.org.uk/project/tight-roaring-circle/>, accessed 9 September 2013.

adults and children together in a recovered *jouissance*, making great claims
for the elastic potential of a mattress:

> Nothing is more fun than jumping on a hard, bouncy bed. All children have enjoyed
> doing this until the mattress collapsed – a bad memory. The double ecstasy of
> the muscular effort in the thighs and calves, a powerful, almost metallic leap, a
> pause in the air that seems eternal, during which the body assumes positions and
> performs. ... Civilization sometimes makes minor progress?[45]

A simple children's game possesses serious political meaning. For Serres,
returning to a state of levity recaptures a vanishing empirical world that
has been lost in the accretions of commodity and regulation. *Tight Roaring
Circle* and *Sacrilege* change our co-ordination with these spaces, associated
with industrialization and the heritage industry, in order to mediate and
reinvigorate the social bonds and participation between bodies and citizens.

Gravity and Grace

Gravity is a medium for the funambulist, climber or aerialist, an invisible
force, which allows the performer to make something out of nothing. It
appears that in these and other examples to be discussed, the circus act
has accrued new value and status within contemporary art, though this
trajectory started at the turn of the twentieth century with the circus act
propelling its speed, stunts, abrupt halts, and vertigo-inducing accelerations
into the modern age of cinema. Circus was a powerful symbol of escape
from the restrictions of modern society, a temporary and migratory space,
whilst its acrobats and clowns provided a mute drama freed from narrative,
their falls and somersaults describing a kind of antic physics. However, the
circus performance was also instituted into modern philosophical discourse

---

45   Michel Serres, *The Five Senses: A Philosophy of Mingled Bodies*, tr. Margaret Sankey
      and Peter Cowley (London: Continuum Books, 2008), 316.

with Nietzsche's tightrope walker in *Thus Spoke Zarathustra*.[46] Circus acts also make a brief and perhaps unlikely appearance in Theodor Adorno's *Aesthetic Theory*, published posthumously in 1970. Drawing on Thomas Mann's allusion to the artist as prankster, Adorno sees a sleight-of-hand, magical transfer between the circus trick and art:

> Thus the once disdained concept of the 'artiste' recovers its dignity. That trick is no primitive form of art and no aberration or degeneration but art's secret, a secret that it keeps only to give it away at the end. Technological as well as aesthetic analyses become fruitful when they comprehend the *tour de force* in works. At the highest level of form the detested circus act is reenacted: the defeat of gravity, the manifest absurdity of the circus – Why all the effort? – is *in nuce* the aesthetic enigma.[47]

Both art and circus centre on an unresolved dichotomy of lightness and weight or object and nothingness. A striking disproportion exists between the energy expended and the sudden illusion of weightlessness. Thus, the body that defies gravity in performance exists as a direct physical model of all that is mysterious in art. A number of issues arise from Adorno's analysis. How does art relate to the effort of work? How does it avoid reification into a commodity? And how does it resist complete or radical dematerialization? How does art also remain tentatively complete? Wire walking can be seen as a model of art's balancing act or precarious identity so that the invisible force of gravity becomes a medium, through which art appears in fleeting manifestations. In its culminating moment of triumph, the *tour de force* is the instant of 'look no hands', a effervescent moment, in which the artist or performer disappears, their own mediation vanishing.

In light of Adorno's 'art as trick' formula, Klein's defiance of gravity might be viewed as the very manifestation of art's precarious identity, its dangerous wayfaring between dematerialization and reification in an emergent period of late-stage capitalism. His trickery of photomontage in *Leap into the Void* describes precisely (whether intentionally or not) the

46 Friedrich Nietzsche, *Thus Spoke Zarathustra: A Book for Everyone and No One*, tr. R. J. Hollingdale (Harmondsworth: Penguin, 2003), 43–4.

47 Theodor Adorno, tr. Robert Hullot-Kentor, *Aesthetic Theory* (London: Continuum Impacts, 2004), 244.

fraudulence at the heart of commodity spectacle, furthered by his clever marketing of the performance. By situating the leap within an ordinary suburban street, Klein seems to betray his own aspiration for a universal art and architecture of air, which by its very nature cannot be located in a precise here and now or within the tired finitude of commodity culture and conventional urban space.

Adorno's aesthetic understanding of the enigmatic *tour de force* shows how art might be better understood in relation to the circus performance. Pasquette's trembling prevarication on the wire points to the problem of art's condition of the cliff edge, neither pure matter, nor pure spirit. It also points to its truth-telling potential as authentic experience in its existential danger. Modern artists appropriated the image and sign of circus because it was indicative of the mysterious operation of art's illusory power, but also the circus performer exuded pathos by virtue of his or her social marginalization. Nietzsche had already captured this existentialism in the image of the tightrope walker, imagining a reinvigorated life through the aerialist's performance of nerve-shattering skill. The tightrope walker is threatened by the fall in his 'dangerous trembling and halting', but his fear and prevarication on the rope mark an eloquent though calm acceptance of death.[48]

A relationship between circus performer and artist existed on the felt assumption that both were excluded or self-excluded from mainstream bourgeois experience. Of course, circus was no more immune from capitalist ideology than anything else; in fact, Chaplin's mechanical worker in *Modern Times* (1936) recoups the tragic history of the circus performer as Taylorist dream. Circus merely offered the sign of freedom with its migratory operation, liminal space, floating bodies, and otherness. Even the instinct or appetite for danger was drained from the performer's body through constant repetition and rote action. Like Walter Benjamin's prostitute, the circus *artiste* is both seller and commodity, alienated as both artist and worker. Pasquette's high-wire performance preserves the ghost of the circus act. Drawing across the sky, Yass echoes the abstract tendency in art wherein the performing body was co-opted into a pure language of

48   Nietzsche, *Thus Spoke Zarathustra*, 43–4.

form, space, motion and line by artists such as László Moholy-Nagy or Alexander Calder. According to Jacques Rancière, the fascination with the mute mechanics of circus performers and their later incarnations, the film clowns, is not simply a case of bourgeois degradation in low popular forms, but indicates a powerful rationality connecting modern aesthetics, commerce and a fusion between art and life within new 'regimes of perception'.[49] Streamlining, abbreviation and physical lightness combine in an ideal aesthetic equivalence that permeates modern advertising, design, poetry, and art to produce a vision of life as one of common purpose.

High-rise construction is both a pragmatic design solution and a discourse of the abyss-fall, these two thought paradigms reflecting, on the one hand, an abstraction of the object-building produced in flat plan and, on the other, the social production of that space. Building high means combining the dream plan with the potential of a real fall. W. G. Sebald writes in *Austerlitz*: 'Somehow we know by instinct that outsize buildings cast the shadow of their own destruction before, and are designed from the first with an eye to their later existence as ruins.'[50] Whether department store, corporate building or high-rise social housing, modernist projects of building into the sky have made the terror of falling into a potential destiny for the urban citizen.

Why are these gravity-defying climbs and balancing acts on high buildings considered subversive by performer and audience alike? Is it because the physical work they do to move or stay up in the air has no other motivation than play or poetry, converting values of exchange into those of self-determining energy? Skyscrapers, perhaps the greatest representatives of capitalism, are themselves the ultimate sublimation of work. Therefore, play allows for an oppositional strategy to the Protestant work ethic and to those paternalistic modernizers, who moved large populations into the sky on the periphery of cities such as Glasgow during the 1950s and 1960s. There is a lesson for art in Tschumi's notion of cross-programming in architecture. Is the flirtatious overlap between artist and the circus *artiste* part of

---

49   Jacques Rancière, *Aisthesis: Scenes from the Aesthetic Regime of Art* , tr. Zakir Paul (London and New York: Verso, 2013), xii.
50   W. G. Sebald, *Austerlitz*, tr. Anthea Bell (New York: Random House, 2001), 19.

an extended questioning into what constitutes the artwork or *Kunststück*? By using the body to contaminate architecture with burlesque or antic performances, artists have tested other philosophical and metaphysical tenets.

All the contemporary artists discussed in this chapter have acknowledged or challenged gravity by disrupting the structures that give it visible form. This force-sense has become the lens through which artists have approached the questions and conditions of modern architecture and hence modern life. Echoing the 1960s' inflatable moment, Saraceno and Deller use the elastic potential of plastic and air to stretch or prolong the fall, activating pleasure and pop emancipation. Art strays into the domain of architecture in order to respond to the real imperatives of modernity and capitalism by witnessing the spectacle of its terminal condition. Fascination with the post-war architecture of housing estates in the work of Yass, Whiteread and Gaillard reflects the potent dialectic of catastrophe and spectacle. Why does failure need to be so graphically relayed in the spectacle of falling towers? Is the answer found in the way catastrophic falling and the utopian longing to fly have converged in fantasies of collective belonging or solitary dreaming, meeting at a limit, at an aesthetics of the cliff edge? Like the void or the air, gravity is an invisible medium through which art and its unacknowledged cousin, the circus trick, appear and then disappear from view. The tightrope walker is a recollection of the philosophical and aesthetic condition of modernism, its 'dangerous wayfaring'. The position of art since the 1960s is one of delicate equipoise, of neither dematerializing, nor reifying. Cast into an *aporia*, art increasingly resembles the quality and metaphor of the high-wire circus act. It must exist in a state of constant displacement, never quite taking root or setting itself up on firm foundations. Like the gravity-defying, risk-taking spontaneity of the circus trick, art continues to step out over the abyss and live on shaky ground.

CHAPTER 2

The Cosmic Cage

> Who dares to think you can play with matter, that you can shape it for a joke …[1]

Gravity is a sliding door between everyday experience and mysterious force, between the simple acts of getting up in the morning to recent advances in the understanding of that particle that gives matter, mass, the Higgs boson.[2] My subject in this chapter is how contemporary artists have navigated the space between the subjective experience of our basic force-sense, gravity, and its inexplicable, occult quality, unconsciously shadowing the scientific thought experiment. In contrast to assaulting or disrupting the orthogonal rectitude of architecture in its Newtonian law-giving variant, the artists in this chapter play with understandings of gravity as a force of dark agency, whilst examining assumptions about human will and agency. The artist most explicitly seized by the phenomenon of gravity, Bas Jan Ader, reflected on profound questions of its relationship to causality, metaphysics, ethics, and will by submitting to gravity's mastery in a series of falling events until his early death in 1975.

There is a counter-intuitive doubt abroad in the following artworks and performances, by which things, objects, and matter play with the artist,

1 Bruno Schulz, *The Street of Crocodiles*, tr. Celina Wieniewska (New York: Penguin, 1977), 64.
2 Higgs boson particle was confirmed after a long theoretical hypothesis in 2013 at CERN.

rather than the reverse, in the manner of Montaigne's cat.[3] By imparting the effects of gravity to inanimate objects and enlivening inert matter, artists such as Fischli and Weiss, Wood and Harrison, Rodney Graham, and Richard Wentworth often delegate agency from themselves onto other objects and forces in the manner of *deus ex machina*. As if attributing matter's vitality to the mysticisms of religion or magic, artists have played with the notion of a 'prime mover' or hidden cause by simply engaging the agency of gravity or by invoking the magic act to refract these residues of belief empirically. Departing from an anthropocentric or monotheistic view of the world, these contemporary artists have reshaped the idea of falling as willed intention to one of mishap, enigma, or cosmic joke, enlisting the active participation of gravity in ad-hoc assemblages between body and object. Gravity connects to deep considerations of matter's weightiness, possessing the potential to move or reshape our habits of understanding in its invisible and obtuse reality. What characterizes works by Wentworth or Fischli and Weiss, as so many artists from the 1950s such as Rauschenberg and Oldenburg, is a propensity to revalue waste materials, everyday objects, and obsolescent technologies through creative play. The artist is engrossed in tinkering and assembling, recalibrating relationships with the object. As a result, the studio space merges with environments of garden shed, theatrical prop cupboard, science laboratory, or garage as a place for play, experimentation, and performance. Within these spaces and fields of force, things and bodies stick together, collide, fall, or join up.

I explore the artworks and performances via more recent theories of 'new materialism'. More recent scholarship, characterized as 'object' or 'new materialism' revisits older materialist traditions such as vitalism and animism, but with a reinvigorated philosophical purpose that departs from anthropocentric certainty and attributes greater agency and independence to non-human objects or forces. New materialists reinvest objects with 'thing power', so that human agency is placed in a more equal standing with objects within a wider network of material formations, redrawing the balance between ourselves and nature. The ecological and ethical implications

3   Michel de Montaigne, 'When I play with my cat, who knows whether I do not make her more sport than she makes me?', *Apology for Raymond Sebond* (1592) <http:// www.sophia-project.org/>, accessed 1 May 2017.

of new materialism seem to move in two key directions, one that drives human beings into states of impersonality, democratically equal to animals and objects; the second which turns objects into eloquent beings, elevating their character. From these considerations, how do ideas of agency and will become significant to the creative energies of artists?

Clown Fall

In the examples that follow, artists navigate the indiscriminate force of gravity through an insistence on the body's 'object-ness', most clearly seized in the act of falling. In terms of how artists have trespassed into the domain of circus, let us turn to the discourse of the clown. Falling is the gerund-state than unites all bodies and objects on earth and the clown's falling is emblematic of this most rudimentary condition. Embodying keynotes of modern dislocation, the clown is physically and emotionally adrift in a world of increasing efficiency and automaticity. Clowns such as Chaplin, Keaton, and Lloyd repeat the fall as a profound condition of being, whilst doing solitary battle with the order of things. Continuously challenged by gravity, the clown reveals the body's mass, rudely mechanical in its propensity to fall. Such alienable effects of the body, its heaviness and inertia, arise in the clown's inability to keep pace with the flow of events and objects that run in his path. In his short 1911 treatise on laughter, Henri Bergson notes the incompatibility between the flow of life and the movements of the clown, exaggeratedly impeded by costume and oversized boots:

> No more in our eyes than a heavy and cumbersome vesture, a kind of irksome ballast which holds down to earth a soul eager to rise aloft. Then the body will become to the soul what, as we have just seen, the garment was to the body itself – inert matter dumped down upon living energy.[4]

---

4   Henri Bergson, *Laughter: An Essay on the Meaning of the Comic*, tr. Cloudesley Brereton and Fred Rothwell (Book Jungle, 2008), 31.

For the clown, the ground provides a constant round of call and response. Summoned by collapse at every turn, the clown's body is connected to gravity like a noose. Humiliated, inverted bodies, repeatedly the subject in Goya's *Disasters of War* (1810–1820), are further objectified by the Chapman Brothers' *Insult to Injury* (2004) series, in which they draw over the Goya etchings with clown faces.

Bruce Nauman's *Clown Torture* (1987) is a four-channel video installation, which loops four different narratives with four different types of clown; a jester, a French baroque clown of the *commedia dell'arte*, a dumb clown, and a traditional polka dot, red-haired show clown, played by Walter Stevens.[5] All of the short sequences are looped, so that there is no sense of a beginning or end, and all of the clowns are tortured through different traps; a falling water bucket, a tricky and never-ending balancing act with a goldfish bowl and a broom handle, the interminable sequence of the 'Pete and Repeat' children's rhyme, and the ground itself. Dressed as the traditional dumb clown, Stevens falls backwards on the floor and kicks his legs, shouting 'No, no, no, no' at some nameless terror as the camera zooms in on his oversized shoes. His continual wailing is somewhat at odds with the normally mute performances of the circus clown, rolling and heaving his body into a state of silent object-hood. Nauman does not explain the clown's ordeal, but simply entraps him within a small room without props, and within an endless cycle of apparent torment or existential pain. That the clown never gets to the point of the joke or the punch line is itself the allegory of existence, of life lived in an unending cycle of attempted levitation and repeated collapse.

A determining feature of Nauman's work is his acknowledged debt to Samuel Beckett, with a one-step conversion from tramp to clown. In his 1968 video work, *Slow Angle Walk (Beckett Walk)*, the artist turns a fixed camera on its side, recording 60 minutes of awkward walking based on the tortuous gait of Beckett's 'Watt', with legs raised up at right angles to the body both in front and behind. With the camera turned on its side, the ground

---

5   'Bruce Nauman, Clown Torture', Art Institute Chicago (1987), <http://www.artic. edu/aic/collections/artwork/146989>, accessed 19 August 2013.

becomes the wall, challenging normative readings of gravitational space. As he paces out his ritual walk, he complicates the easy assumption of how we resist the ground.[6] In an agonisingly slow walk, Nauman holds his body in awkward states of suspension, trapped in a futile circuit of movement.

Duration and endurance are synonymous in Nauman's early performances, as the rituals and repetitions become patterned, mantra-like, into his body. Through iterating the ground in different formations, Nauman distributes agency to the ground in a more equal *pas de deux*. In *Stamping in the Studio* (1968), his whole body disappears off camera with only his footfalls still audible. In *Bouncing in the Corner I*, Nauman allows his body to fall against the corner of the room and bounce off the walls with repetitive insistence, conveying its object-like quality and weight. There is no *jouissance*, only the thud of 'here-ness' and 'now-ness', with his body caught in the binary flick-flack of energy transfer from wall to body. The performer is alone, reassured only by the dull echo of wall or floor, physically objectified by the room's reflections. Nauman registers Beckett's grim evocation of Bishop Berkeley's theological premise, 'to be is to be perceived' (*esse et precipi*), so that the sound of a ball bouncing, or even the fall of the foot in the empty studio, returns a fragile sense of the self.

Double Act

Beckett's tramp figure recalls the clowns and tramps of circus, music hall, vaudeville, and silent film. The popular pathos of songs such as *Underneath the Arches* sung by the music hall act, 'Flanagan and Allen', the blank face of Keaton, who took the main role in Beckett's only film (*Film*, 1961) and the funny walks of Chaplin and Max Wall were formative for Beckett, and variously inscribed throughout his plays and prose. Krapp (*Krapp's Last Tape*, 1958) is represented with clown-like features: he has a white face, purple

---

6   Steven Connor, 'Shifting Ground' (2004) <http://stevenconnor.com/beckettnauman.html>, accessed 23 January 2017.

nose and wild hair and suffers with his bowels. In the opening of the play, Krapp, eats the quintessential comedy fruit, a banana, and nearly slips on its skin. In *Waiting for Godot* (1957), Vladimir (played by Max Wall in a 1975 production at Greenwich Theatre) and Estragon perform a quick-fire three-hats, two-heads routine in Act 2, mimicking a scene from the Marx Brothers' film, *Duck Soup* (1933), and the double-act comic patter of music hall. Like the Vladimir-Estragon dyad, characters are paired in hellish routines and habits of manic crosstalk: Hamm and Clov, Nagg and Nell (*Endgame*, 1957), Winnie and Willie (*Happy Days*, 1961). Beckett alludes to Bim and Bom in many of his works (sometimes called Bom and Bem), a Russian clown duo of the 1920s, who were sometimes known by the singular name, *Bim Bom* in strangely conjoined agency. In the prosthetic condition of double agency, Beckett's characters submit to the plug-in, add-on auxiliaries of other people and objects. Hamm is immobile, confined to a wheelchair, but Clov cannot sit down; Vladimir is heavy and Estragon, light; Willie can crawl, but Winnie is buried in earth, first up to her waist and in Act 2, up to her neck. With bleak irony, Winnie is absorbed by the thought of escaping gravity:

> WINNIE: [Long pause.] Is gravity what it was, Willie, I fancy not. [Pause.] Yes, the feeling more and more that if I were not held – [gesture] – in this way, I would simply float up into the blue. And that perhaps some day the earth will yield and let me go, the pull is so great, yes, crack all round me and let me out.[7]

This is mirrored by the exchange between Vladimir and Estragon in Act 1 (*Waiting for Godot*) in which Estragon asks whether they are 'tied' down to Godot just before Lucky enters physically bound by a rope to his master, Pozzo, and struggles with the weight of his baggage. Gravity, grey light, and a bleak horizon are often the minimal resources to indicate the co-ordinates of where the 'so-called' action takes place; the ground a condition of 'last resort'.[8] Moved beyond the confines of a monotonous space, characters, such as 'Murphy' dream of peace within a limitless space, free of gravity:

---

7   Samuel Beckett, *The Complete Dramatic Works* (London: Faber and Faber Ltd, 1986), 151–2.

8   Connor, 'Shifting Ground'.

'But how much more pleasant was the sensation of being a missile without provenance or target, caught up in a tumult of non-Newtonian motion.'[9] This limbo state is reflected in Beckett's critical use of light in so many of his plays, reflecting his interest in Gnosticism, and in particular, the major Gnostic religion, Manichaeism. Its dualistic cosmology was based on an understanding of the world as essentially fallen or perverse, a cosmic error in which the force of good, represented by light, was overcome by the force of evil or darkness. John Calder confirms Beckett's valorization of Manichaeism in his writing:

> In so far as Manichaeism is concerned there is much to say, not only of his direct fascination with one of the most interesting of the so-called heresies, but in the recurrence of opposing twosomes throughout his writing. His characters tend to come in pairs, Vladimir and Estragon, Pozzo and Lucky, Hamm and Clov, Nagg and Nell, Mercier and Camier, Molloy and Moran and so on, each tending to be the logical opposite of each other. Manichaeism contrasts good and evil like day and night and is based on a perception of the world that, while admitting that God might have created it, also recognizes that he did so very ineptly, and probably did not quite realize what he had done. He might therefore have appointed a deputy to run things for him, which to the Manichaeans could well have been the Devil.[10]

After reading Arnold Geulincx's *Ethica* (1675) in the 1930s, Beckett accommodates the phrase, 'where you are worth nothing, you should want nothing' into his writing.[11] Geulincx makes a radical argument for the strict division between body and mind, regarding the body as someone else's affair and impossible to control. If the body is wholly contemptible and not subservient to the will, he argues, then this necessitates a renunciation of the will and a radical humility, served by deep introspection. As is so often the case in Beckett, the body is brought low, made abject, and shown to be worth nothing. He drains the body of energy and passion, reflecting Geulincx's reading of the erroneous body as a site of unruly drives and

9   Samuel Beckett, *Murphy* (London: Faber and Faber Ltd, 2009), 72.

10   John Calder, *The Theology of Samuel Beckett* (Richmond: Alma Classic Ltd, 2012), 11–12.

11   Shane Weller, 'Beckett and Ethics' in S. E. Gonstaski, *A Companion to Samuel Beckett* (Chichester: Wiley-Blackwell), 125.

deviation. Introspection, the tactic favoured by Murphy, is connected to the demands of seeking *gnosis*. As John Calder confirms, '[Beckett] had also connected the Gnostics with Geulincx, used Manicheism as the foundation of *Krapp's Last Tape* and followed the philosophy of Schopenhauer who can be connected to Geulincx...'[12] Gravity and heaviness are reflected in the varying states of mobility or rigidity of Beckett's characters, mirrored also in the diurnal nightmare of a world of dull mass unleavened by light or lightness. Gnostic doubt took the idea that God existed, but that man was distant from, or ignorant of the real deity. As a distant or buried memory, the grace or knowledge of God might be activated by the search for inner enlightenment. In the Gnostic tradition, the world is a stark and alienating place for human beings, simply the bungled creation of an evil demiurge.

Philosopher, Simone Weil, whose aphorisms were collected into a posthumous publication, *Gravity and Grace* (1947), practised ascetic self-denial, symptomatic of her philosophical project. Suffering was to be that form of violence turned against the body, a way of making it 'other' or alien. In his preface to Weil's *The Need for Roots* (1943), T. S. Eliot branded her a modern-day Gnostic or Marcionite in light of her idiosyncratic handling of religion.[13] For Weil, gravity reveals the condition of mankind and the absence of God; gravity is the symbol of the law, which puts force on the side of baseness, this force being violence, cruelty, and amorality.[14] Nevertheless, the grace of God is possible with sufficient humility. In the Gnostic vision, human beings are condemned to a life without transcendence or hope on earth, unless inner enlightenment is practised and found. The connections with gravity are highly relevant. Instead of light or spirit conveying the possibility of transcendence, light is only present to expose a painful vision of heavy matter and stranded bodies, lacking meaning or trajectory. For the Gnostics, reproduction was a duplication of this absurd joke. Beckett's use

---

12   Calder, *Theology of Samuel Beckett*, 54.

13   Jeffrey Mehlman, *Emigré New York: French Intellectuals in Wartime Manhattan 1940–1944* (Baltimore, MD: The John Hopkins University Press, 2000), 87.

14   Simone Weil, *Gravity & Grace*, tr. Emma Craufurd (London and New York: Routledge, 1997), 2–4.

of a grey half-light expresses this purgatorial suspension in a world drained of true light or spirit. Beckett echoes this thought in *Murphy*, who perceives nothingness and colourlessness as 'a rare postnatal treat'.[15]

As stated, Beckett's double acts, most famously Vladimir and Estragon, were adopted from the popular traditions of music hall and vaudeville, as well as circus and screen. There has been a rich tradition of the double act in British light entertainment, which has filtered through to performance art. In their radical disruption of the formalist sculpture of Anthony Caro at St Martin's College of Art in the 1970s, Gilbert and George have used their bodies, attitudes, and everyday rituals to break the fixed parameters of sculpture. Part of their guise is the English gentleman's suit, used to test questions of identity and sexuality. In *Magazine Sculpture, Two Text Pages Describing Our Position*, a work commissioned by *The Sunday Times* in 1970, they are photographed leaning against a country gate and in the accompanying text invite the viewer to look closely at certain similarities and anomalies. Their stance of relaxed physical pose is reminiscent of Gainsborough's *Mr and Mrs Andrews* (c. 1750) with its Rococo easiness. As the accompanying text points out, viewers should pay particular attention to the hidden geometry: 'Think of all that diagonal relaxation, for only the picture behind is symmetrical'.[16] In his 1952 article, 'The Upright Posture', Erwin Staus examines how bending and inclination 'first brings us closer to another. Inclination, just like leaning, means literally 'bending out' from the austere vertical.'[17] Gilbert and George appropriate the image of the gentleman-tramp duo in order to question the foundations of solo subjectivity and identity. By appropriating the double act, the artists divide agency between themselves in terms of making their work. Cleaving more generally to an almost robotic uprightness, Gilbert and George's use of diagonal leaning is subversive. As Steve Connor sees it: 'The diagonal has always been enigmatic and rather suspicious (the shifty look, the bend

15   Beckett, *Murphy*, 154.
16   Daniel Birnbaum and Jochen Volz, eds, *Making Worlds* (*Fare Mondi*, 53rd Venice Biennale Catalogue), 2009. MOMA owns a new 2017 gelatin silver print of the original work.
17   Erwin Staus, 'The Upright Posture', *Psychiatric Quarterly*, 26 (1952), 529–61 (539).

sinister of illegitimacy). It is the vehicle and dimension of the incalculable, the infinitesimal, the asymptotic.'[18] Bending as both inclination and bodily gesture is involved in a number of their works, for example, their 1976 performance, *Bend It* from the single by *Dave, Dee, Dozy, Beaky, Mick & Tic* of 1968, in which they dance to the music with deadpan enthusiasm and intensity, leaning in and out as they move to the music.

This performance recalls aspects of Gilbert and George's early triumph, *Singing Sculpture* (1969), a parody of the down-at-heel gents' routine of music hall stars, 'Flanagan and Allen', tramps reduced to living underneath the arches yet able to dream and sing. Similarly, music hall double acts based on the hobo clown figure were of fundamental significance in Beckett's *Waiting for Godot*. Painting their faces bronze and moving slowly to the music, Gilbert and George animate and reverse the tradition of sculpture as inert material, at the same time exploring the body as sculptural material. The hypnotic effect of the early 8-hour performances of *Singing Sculpture* astounded the artists themselves with George wondering why a table, a walking stick, and a mono record held such magical effect?

Their double act signalled a radical change in the idea of agency as emanating from a single artist. Double acts are more common in contemporary art these days: Elmgreen and Dragset, Fischli and Weiss, Allora and Calzadilla, the Chapman Brothers, Langlands and Bell, Noble and Webster, and Wood and Harrison are prominent names amongst others, fragmenting the notion of individual authorship. In 2016, artist Richard Wilson co-ordinated an exhibition at the Royal Academy's Summer Exhibition around the theme of artist duos (*Seeing Double*), partly in response to the Royal Academy rule that joint membership is not permitted.[19] The British comedy double act has enjoyed particular success with the enduring popularity of *Morecombe and Wise* or the *Two Ronnies*. There is a way in which it has given freedom to the idea of male togetherness and tenderness, binding audiences to this image. There is poignancy in these images

---

18   Steven Connor, 'Inclining to the View' (2010) <http://stevenconnor.com/inclining.html>, accessed 4 January 2017.

19   Fiona Maddocks, 'Double Vision' (19 May 2016) <http://www.royalacademy.org.uk/article/video-jake-chapman-studio>, accessed 1 April 2017.

of co-dependency with the comic double act operating as a ready-made of man's prosthetic and pathetic condition.

This seam of British vaudeville and the comedy double act is resonant in the work of John Wood and Paul Harrison. Samuel Beckett's play, *Waiting for Godot* has also been cited as a source for the work of this British artistic duo with their focus on the painful symbiosis between two people, and their staging of short, space-time accidents. Where Beckett abstracted the tortuous dynamics and crosstalk of the comic double act from music hall, Wood and Harrison insert the dyadic nightmare of Vladimir and Estragon or Pozzo and Lucky into Minimalism's neutral language of cubes, boxes and architectural-sculptural containers. Acting in tandem, they rehearse a series of silent falling scenarios, which not only serve to corrupt the purity of the white cube aesthetic, but also express the dogged and deadpan determination of stand-up routines.

In *Twenty-Six (Drawing and Falling Things)* (2001), the duo's references to minimalism are contained within the method of filming with a

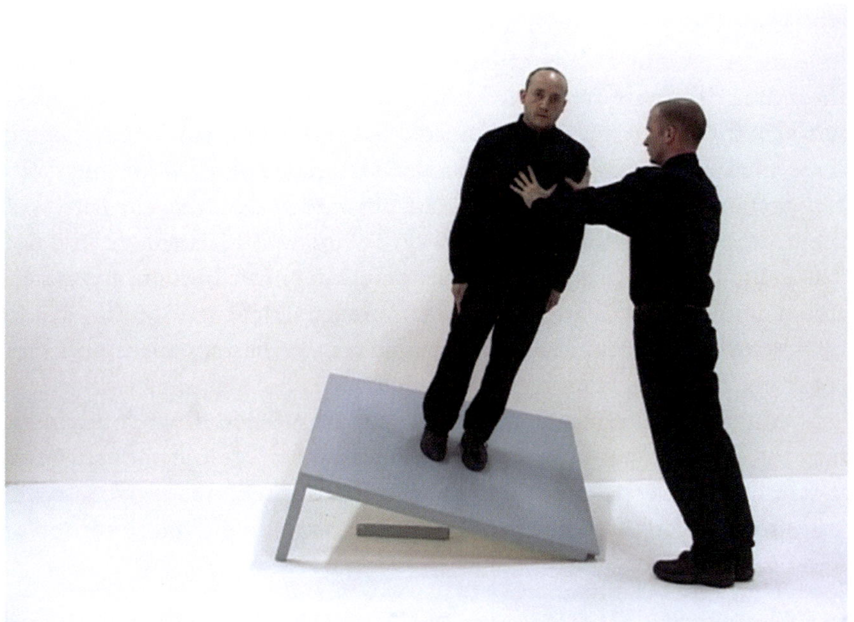

Figure 6. John Wood and Paul Harrison, *Twenty-Six (Drawing and Falling Things)*, 2001.

steady camera of fixed viewpoint, even lighting, and careful framing. In this way, the 'Laurel and Hardy' combo of Wood and Harrison reproduce the proscenium arch of the stage, but dress it in a simple, pared-down aesthetic. Here, the artists work from sketches and diagrams, constructing the architectural containers that frame the performances, and then film each vignette from a fixed viewpoint. In one piece from the series, Wood (the shorter of the duo) stands on a low table and his partner kicks one leg out. Wood then slides down the broken corner and Harrison catches him. Maintaining an utterly neutral look with white backgrounds and blank expressions, Wood is turned into an object, equal with any other falling object. In another short tableau, Wood falls through a diagonal funnel, which is filmed in cross-section. In another work from the series, both artists stand on a rocking semi circular structure, again filmed from side on. In this instance, there is a suggestion of co-operation and trust rather than a nightmare master-slave dynamic. These short scenes are all gag and no narrative, lacking rational motivation or causality. Often the resources are simply the camera, human figure, object or architectural container and gravity. Whilst it is the artists who have planned and configured the outcome of each piece, the audience is made to feel conscious that some external force is dictating the outcome. Wood and Harrison rehearse the rituals, habits, and needs that exist between people, staging these small insights through a pragmatic language of the body and splitting agency between both artists. David Batchelor describes the pathos of their interaction in the following way: 'Wood and Harrison expose not just their "bodies" vulnerability but something more hidden, an orchestrated emotional frailty as well as a kind mutual dependence that would be destroyed if articulated in words. It's very male, very silent and very touching.'[20]

Gravity alienates the body as the site of subjectivity, converting it into an object of weight. As Bergson's reading of the comic body suggests, it is often the inertia of the body within the context of a moving world that provokes laughter. Inertia is both a physical and mental condition of

20   Charles Esche and David Batchelor, *John Wood and Paul Harrison* (London: Ellipsis, 2000), 8.

powerlessness and loss of agency as if gravity were overcoming the will to move. In Beckett's *All that Falls* (1957), Mrs Rooney drags her feet continually and complains, 'Oh let me just flop down flat on the road like a big fat jelly out of a bowl and never move again!'[21] Gravity is the invisible canvas for Wood and Harrison's experiments in balancing and falling, revealing the human body as just another object, democratically conditioned by gravity's downward force. In one experiment, the artists sit on office chairs moving on castors in the back of a typical white van as it moves, crashing into each other, helpless to counteract the movement of the vehicle. Tension builds as the van's acceleration and deceleration spin the artists round with their faces remaining inert in spite of their bodies moving like colliding particles. As in the work of Bruce Nauman, this Sisyphean quality of rugged insistence to keep going is replicated in the way each experiment ends as a type of ordeal. By converting the body into the condition of a billiard ball, the artists produce an image of perilous vulnerability.[22]

Another artist deeply influenced by Samuel Beckett is the Canadian artist Stan Douglas. Like Beckett, Douglas explores the Gnostic vision, particularly in relation to Manichaeism, a religion deeply connected to Gnosticism in its rejection of the physical world and the body. As in the Gnostic tradition, the life of matter was seen as essentially sinful and ruled by a despotic demiurge or trickster. In its radical dualism, Mani's cosmology was based on the primal struggle between light and dark so that light is imprisoned within the human body, as sin descends into the world of matter. This cosmology of light and dark is refracted through black and white photography in Douglas's *Midcentury Studio* of 2011. This series of forty restaged black and white photographs is based on real source material, but framed as the work of a fictional Vancouver photographer, productive between 1945 and 1951. Douglas uses actors, props, and lighting in the manner of a film director to create these photographs, capturing not only a period look but re-examining underlying social and political tensions from less than a lifetime ago. He explores the worlds of crime, gambling, fashion, sport, and the circus, revisiting them with loose allusion to the

21   Beckett, *Complete Dramatic Works*, 174.
22   John Wood was actually injured in the van performance.

press photography of Ray Munro and Arthur Fellig, known as 'Weegee' (who had a spooky ability to time his arrival at a crime scene just the right moment). In Douglas's use of darkness and light, there is a nod to *film noir* with its *topos* of gambling dens and crime scenes; however, as Pablo Sigg notes in his catalogue essay, the cosmology of Mani is also configured through the photographs.[23] Philosopher, Vilém Flusser relates the ontology of black and white photography to the Gnostic philosophy of Manichaeism:

> Black-and-white photographs belong to the same sort of manicheism, only they involve the use of cameras. And they too actually function: They translate a theory of optics into an image and thereby put a magic spell on this theory and re-encode theoretical concepts like 'black' and 'white' into states of things.[24]

Douglas brings together both registers of light; one reflecting the historical use of lighting in this style of photography, and the more disturbing Manicheistic meaning of light as that which flees from the dark matter of this world. Light is a momentary phenomenon that illuminates the world of cosmic error and signals the fracturing of historical time. Artist Pablo Sigg establishes a persuasively argued reading of Douglas's series in relation to Gnosticism:

> The Gnostics – and in particular the disciples of Mani – deplored history as a falsely coherent illusion of time. They felt that time in its totality only involved three moments that somehow split eternity into two: the time prior to Light's encounter with Darkness, the moment of that encounter, and the corruptible time following it, or the descent into matter.[25]

Darkness is therefore synonymous with the fall into gross matter and sin. In the context of Douglas's photographs, darkness not only represents mass, it is also a negative reading of space as a void. In *Shoes, 1947* (2010), the photograph reveals only the lower half of a woman's body with light angled towards her legs and feet. The high-heeled shoes she wears are too large

---

23   Pablo Sigg in Tommy Simoens, ed., *Stan Douglas: Midcentury Studio* (Bruges: Ludion, 2011), 22–31.

24   Vilém Flusser, *Towards a Philosophy of Photography* (London: Reaktion, 2007), 43.

25   Sigg, *Midcentury Studio*, 24.

for her, leaving a gap at the back. Light also illuminates her hands, which are clenched with her thumbs drawn into a fist and catches the metallic strips around the shoes. They gleam with a malevolence that matches her clenched hands, so that the ill-fitting shoes become the mark of an unsettling masquerade. Typical of Douglas's other image dislocations, these shoes are not made for walking, and therefore exude a sinister potential of their own. In *Intrigue, 1948* (2010), a pair of shoes is photographed on top of mirror glass, suspended over a pitch-black background in a topsy-turvy illusion. The reflection is almost more vivid and real than the original with the resulting four shoes twisted in a diabolical dance of ambiguous presence and absence. In *Machine, 1948* (2010), wires are looped within a box, but are apparently defunct or obsolete and in *Hair, 1948* (2010), the elaborate woven chignon, shot from behind, opens up a dark void at its centre. Here the action or modality of weaving subtends a nightmare vision of mass extended without cause or reason. Sigg again:

> Instants like those of the floating dice in *Dice*, the motionless balls in *Clown*, and the sabers [*sic*] flying in a mysteriously symmetrical formation in *Juggler*, seem endless. Like the stroboscopic image *Dancer*, they are extended into pure matter, without time or movement, just as the Gnostics imaged the sensible world, as a kind of material counter-eternity, a vertiginous infinite plasticity.[26]

Weaving is the work done by idle hands, a motif littered throughout *Midcentury Studio* with disembodied hands pointing to small incidents or engaged in light-fingered theft (*Watch, 1950*), as if agency has been disengaged from the individual, issuing instead from a sinister, unseen force. Anonymous hands (like those of the sometimes anonymous photographer) bear witness to supernatural levitations in a macabre reversal of Christian faith, where images often contain representations of saints pointing to the miracles of Christ. Douglas exploits mysterious events and magic acts to emphasize the notion of duplication as diabolical mistake, reflected also in the uncanny of the camera.[27] These strange copies are grotesquely ironic in the way the real subjects of the original photographs have been

26   Sigg, *Midcentury Studio*, 30.
27   Sigg, *Midcentury Studio*, 30.

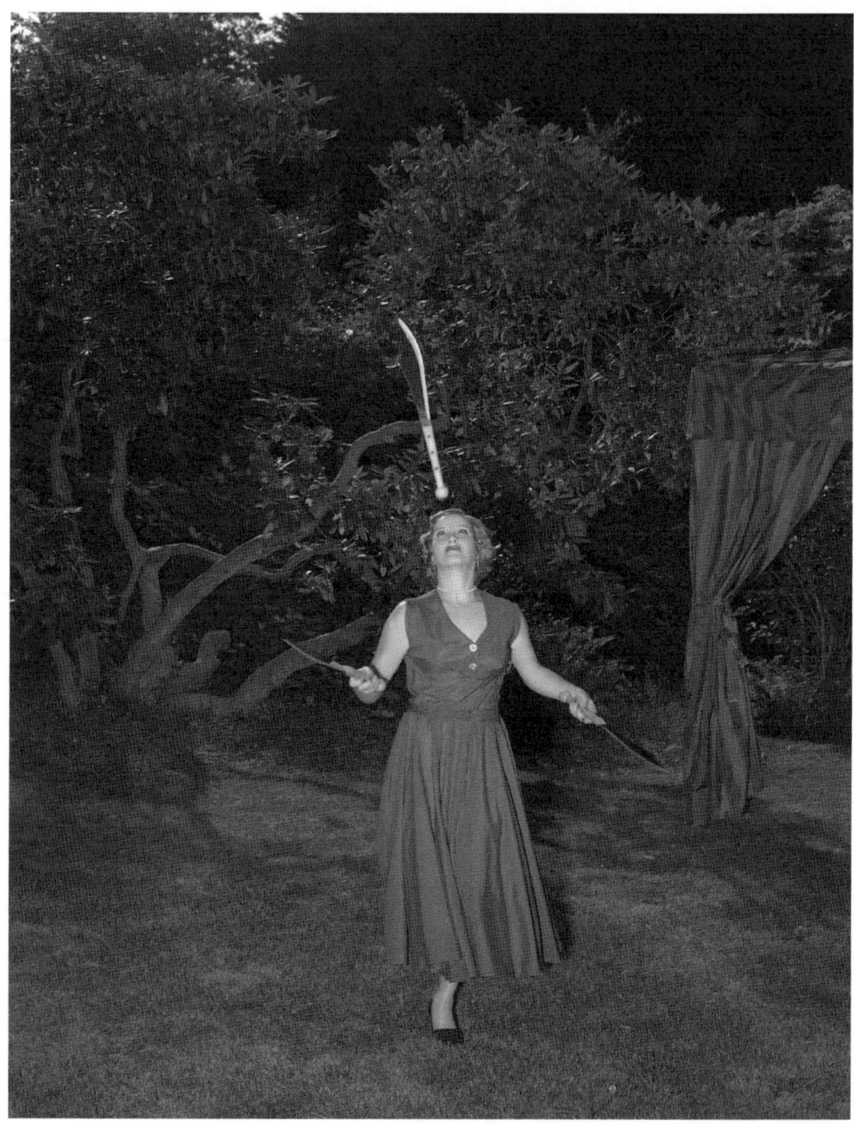

Figure 7. Stan Douglas, *Juggler, 1946*, 2010.

resurrected and reproduced, also mirrored at the image level in the levitations of smoke, dice, rings, and balls, where hidden hands seem to suspend gravity's force. Here the instant light of the camera's flashbulb exposes the world as inherently chaotic, operated by the hand of some invisible, malevolent demiurge.[28] Floating dice or balls indicate a terrifying vision of pure matter, extended and duplicated without meaning in a *mise en abyme*. Photography does not release the subject from destiny and death, but simply reveals the corrupt joke of bodies stranded on Beckett's 'old muckball'.

There is a return to the discourse of clown, in this instance a circus clown, who trickster-like is able to suspend gravity momentarily in a juggling act. For *Clown, 1946* (2010), Douglas draws on an existing photograph of two boys with a Beatty Brothers' circus clown, archived at the Vancouver Public Library. In his restaging, he moves the clown from outside the circus tent into a black photographic studio. The white face and ruff of the clown are sharply illuminated against a black background with one of three oranges suspended in the air as part of juggling act. An expression of fake mirth is plastered across the clown's huge painted mouth and dramatic eye make-up. His hands, which hold the other two oranges, are the only part of the portrait where his real flesh is visible. Ruffs and silky, baggy costume become part of the slippery weave of material and matter, the silk catching the light on its surface.

Douglas's clown is a fraud, a photographic pastiche of a clown, who is already an imposter as indicated by his corrupt disguise. The clown's make-up and glossy costume represent a more general dislocation from original to copy, or presence to absence in the downfall of duplication. As with the floating dice (*Dice*), the orange is suspended in light. Light itself is used as a medium of disguise, a masquerade to deceive the eye, just as juggling was historically an art to deceive the senses. *Clown* retains the meaning of juggling as magical practice, re-invoking in these images, the web of connection between residual magical and the automaticity of technology. In particular, Douglas alludes to an interest in Flusser's

28 Sigg, *Midcentury Studio*, 23.

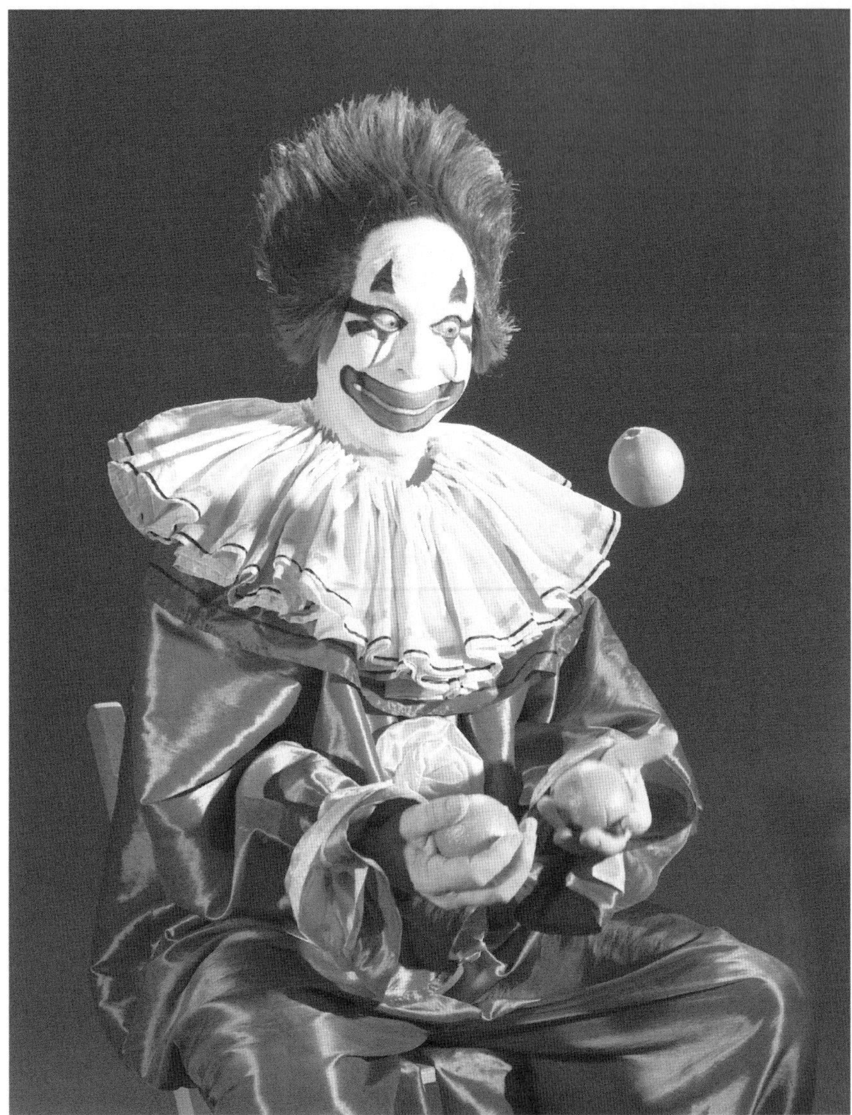

Figure 8. Stan Douglas, *Clown, 1946,* 2010.

negative reading of the 'robotization' of photography by which the camera is already programmed with a certain set of possibilities that do not allow for more creative license.[29] Flusser argues that the camera's automaticity has led to our renunciation of control, delegating power and agency to the camera's technology; as a result, the camera is a black box, which represents our loss of control over the order of images.

> In fact, however, we are manipulated by photographs and programmed to act in a ritual fashion in the service of a feedback mechanism for the benefit of the cameras. Photographs suppress our critical awareness in order to make us forget the mindless absurdity of the process of functionality ...[30]

Flusser argues that the photographic universe returns us to phenomenological doubt, a state in which images proliferate endlessly, but dull the senses. However, Douglas responds by acknowledging this automaticity, whilst speculating that in the instant of the camera's flashlight, certain details or elements cannot be controlled.[31] There is something amiss in these images; shoes that do not fit, a secret hole in a wall, black holes in a hairdo, and a crime committed somewhere close by but off camera. Alongside Douglas's critical appropriation of found photographs, his images of floating dice, knives, or oranges reflect the trick-like aspect of the camera apparatus. As light collects on the surface of these objects, things are imbued with plenitude, density, and weight.

Falling, Falling Again, Falling Better

Bas Jan Ader's examinations of falling, as iterated through his short falling performances of the early to mid-1970s, echo early slapstick comedy as if conceived within a template of Greek tragedy. In contrast to Yves Klein's

29   Phillips, *Midcentury Studio*, 17–19.
30   Flusser, *Philosophy of Photography*, 64.
31   Phillips, *Midcentury Studio*, 19.

muscular and heroic leap into the void, Ader finds a frail and tender spot in his existential disposition to the fall. Emerging from a Dutch Calvinist background, he became interested in Camus's ideas of will and freedom. For Camus, freedom is found in recognizing that we do not possess free will and in accepting our place in the chain of causality. These ideas are manifest in Ader's series of short falling films from the early 1970s, created at a moment when conceptualism was marked by earnestness and logical rigour. Each falling film opens with a situation or predicament; Ader hanging inexplicably from a tree over a river, sitting on a chair on a rooftop, or holding two heavy stones up, over two light bulbs in a dark garage.

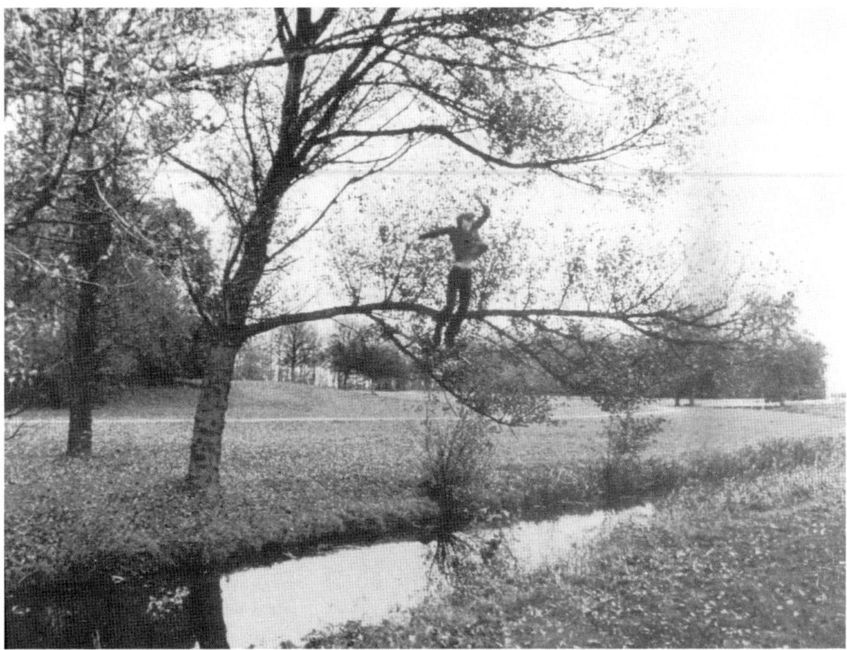

Figure 9. Bas Jan Ader, *Broken Fall (Organic)*, 1971.

Fall II (Amsterdam), made in 1970 is a 19-second silent film in which a static camera records the artist cycling along the *Reguliersgracht* in Amsterdam with a bunch of flowers in his hand, and then veering off into the canal

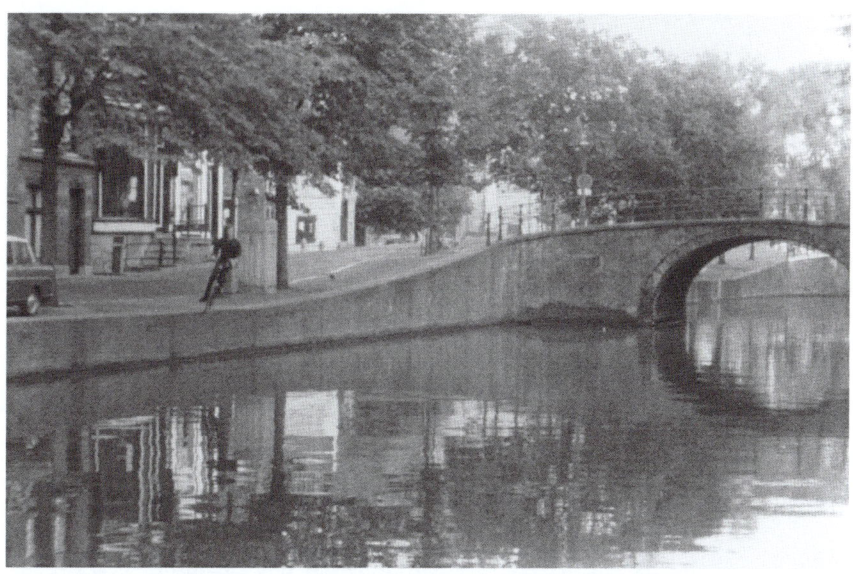

Figure 10. Bas Jan Ader, *Fall II (Amsterdam)*, 1970.

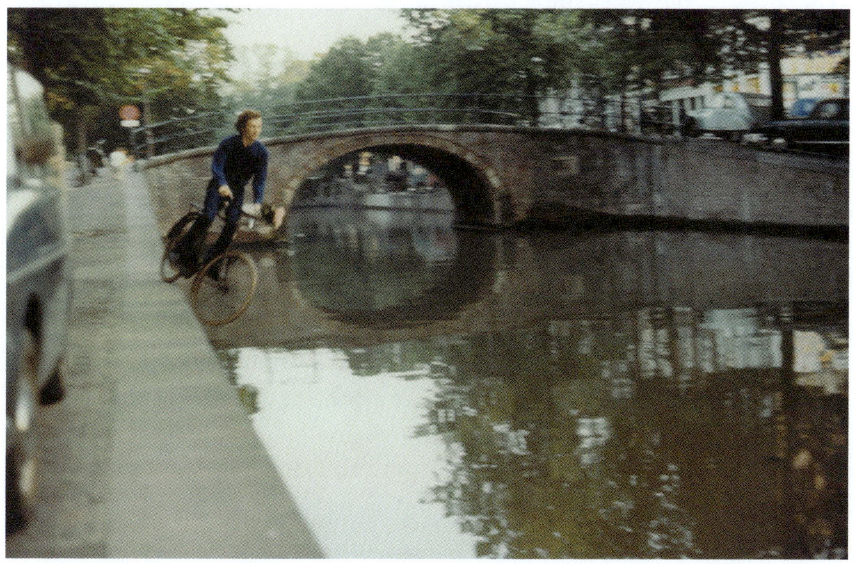

Figure 11. Bas Jan Ader, *Fall II (Amsterdam)*, colour, 1970.

inexplicably. Artists Ger van Elk and William Leavitt were on hand during the performance to photograph the fall. His disappearance into the water recalls Brueghel's *Landscape with the Fall of Icarus* (a 1560s copy of the original) with the hero's tragic fall barely visible in the distant waters. Ader transforms mythic drama into commonplace failure, converting himself from willed body to falling object and dipping into the water, out of the frame and out of view. As the artist announced in 1972: 'I have always been fascinated by the tragic. That is also contained in the act of falling: the fall is failure'.[32] In each fall, there is a short moment of suspense as if the artist were weighing up his options, drawing out time in the way that gravity slows time.

In *Broken Fall (Geometric; blue-yellow-red)* of 1971, Ader references the neo-plasticism of Mondrian, invoking and then subverting the earlier Dutchman's overall rejection of the diagonal. Leaning becomes a critical component in Ader's falls, balancing on a knife-edge between some small residue of will and the inevitability of the fall. The C-type print shows Ader on a brick lane bordered by foliage with the Zeeland Westkapelle lighthouse (famously painted by Mondrian) in the distance. Falling, as a radical event of self-alienation, forms the philosophical heart of Ader's work as an artist, explored as the fundamental basis of understanding the world. He marries logical procedure with the chaos of the fall, pointing each small investigation into gravity towards a larger metaphysics. Ader's investigations into falling are perhaps the most explicit use of gravity as a medium.

Nightfall (1971) is a short black and white film (4′ 8″) set in a domestic garage. Ader stands behind a concrete paving slab and films himself from the front in mid-shot. After a pause, he picks up the heavy slab and lifts it onto his left shoulder, then onto his left hand before dropping it onto a light bulb to extinguish the light. This action is repeated on his right side and when the second bulb is extinguished, the scene turns dark signalling the end. This contrast between light and dark echoes Beckett's Manichaeism, particularly the brevity and simplicity of Beckett's 40-second play, *Breathe* (1969). In both, the termination of light is signalled as a fall from consciousness. His earliest falling performance was *Fall I (Los Angeles)* in 1970, when

32   Alexander Dumbadze, *Bas Jan Ader: Death is Elsewhere* (Chicago: University of Chicago Press, 2013), 17.

he climbed onto the rooftop of a suburban house with a chair, leant out of his gravitational centre whilst seated and tumbled downwards off the roof. Although his leaning is wilful, it should be understood through Hannah Arendt's description of the human condition and her diagnosis that we are 'doomed' to freedom.[33] Ader simply states that he has been mastered by gravity, helpless in surrender to its power. Falling formed the nub of his philosophical interests, particularly in connection to the idea of how human will is helpless within the infinite forces of nature. This may only exist in a personal buzz of interpretation, but Ader's testing of the philosophical injunctions of gravity seem to be the artistic augury of French philosopher, Michel Serres's *The Birth of Physics* (1977). In this book, Serres explores the power and daring of classical atomism, redirecting attention to the occult geometry and poetics of the swerve or *clinamen* as that which opposes the deadening world of regulation and classical physics.

According to Michel Serres, falling is a paradoxical condition that is both generative and destructive of order. In *The Birth of Physics*, Serres reclaims the atomism of Lucretius' poem, *On the Nature of Things*, and traces a history of the angle, the swerve or *clinamen*.[34] For Lucretius, the unpredictable swerve occurs at any point in time and space and is generated by atomic weight.

> When atoms are borne straight down through the void by their own weight, at a wholly unknown time and unknown point they swerve a little from their course, just enough that you could say they had changed their direction. Because if they did not, they would fall like drops of rain through the yawning void and no collision of atoms would ever have occurred, no blows been caused: In which case, nature would never have created anything.[35]

This minimum deviation or angled fall is so small that it cannot be seen or measured, but it is the crucial engine of creation. Serres argues that

---

33   Dumbadze, *Bas Jan Ader*, 31.
34   Michel Serres, *The Birth of Physics*, tr. Jack Hawes (Manchester: Clinamen Press, 2000).
35   Philip de May, *Lucretius: Poet and Epicurean* (Cambridge: Cambridge University Press, 2009), 34.

classical physics is a discourse of hard, solid matter and predictable forces as enshrined in Newtonian physics. Such accounts of universal physical laws are by implication forged on the basis of a single, omnipotent creator. By contrast, the atomism of Lucretius promotes an empirical view of creation, one of chance admixture and inexplicable deviation. The study of flow and swerve in atomistic theory produces a far more unpredictable account of how things emerge without a clear point of origin in time and space. By extension, this view of nature places man in the midst of turbulence or unpredictable flow. It weaves the subject into the flux of matter, introducing voluptuous knowledge and empirical sensation into the discourse of science.

It seems that falling is an important conceit in the work of Rodney Graham. Graham's complex allusions to falling are evident in what he names his 'costume film trilogy'. In *Vexation Island* (1997) and *City Self/Country Self* (2000), two films within this trilogy that also includes *How I Became a Ramblin' Man* (1999), falling bodies and objects are the engines of narrative loops in a Freudian hell of eternal return. Graham refers to this repetition function as a structuring element in the film. *Vexation Island* (9 minutes, 20 seconds) is set on a small tropical island. A buccaneer (Graham himself) is asleep on a beach with his head on a barrel and a parrot perched close by. It is unclear whether he is asleep or has been knocked unconscious, because he has a bloody wound on his forehead. In the first 7 minutes or so, Graham mixes a number of different shots of the small island, the beach, the palm tree, the parrot, and the sleeping pirate but without any clear narrative drive or event in contrast to the all-action 'Treasure Island' narrative. Towards the end of the film loop, the pirate wakes up, stands up and looks up into the palm tree, before being hit on the head by a falling coconut. Does the falling coconut echo Newton's falling apple that, by his account, fell on his head and sparked his eureka moment? *Vexation Island* is clearly Edenic, although the coconut – as ersatz apple – is imbued with slapstick associations (like the fake clip clop of a horse in amateur dramatics). Graham juxtaposes the falling object with the falling buccaneer, emphasized through slow motion and conceived as a fall from consciousness. Graham explains his purpose: '*Vexation Island* is a costume picture, that is to say a travesty; but above all, the short film aspires to be an accurate diagram, pleasingly

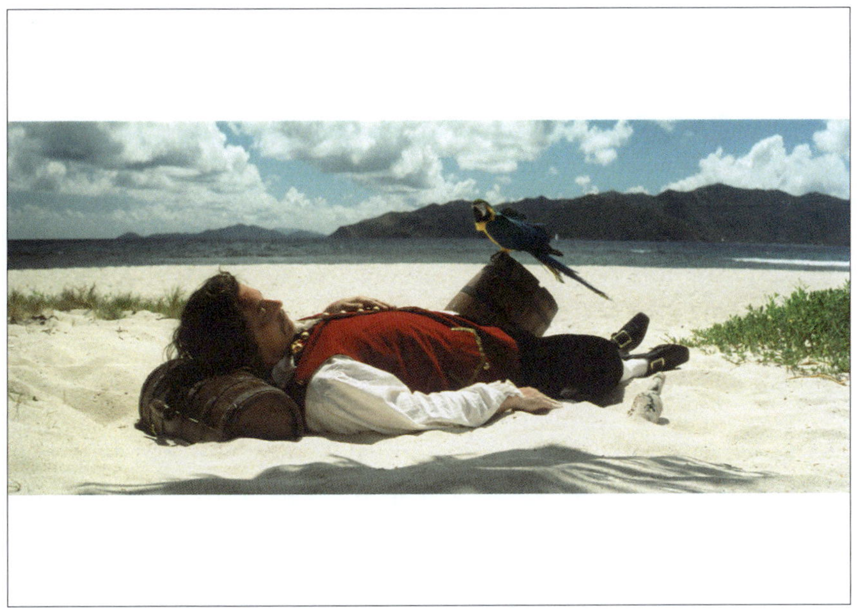

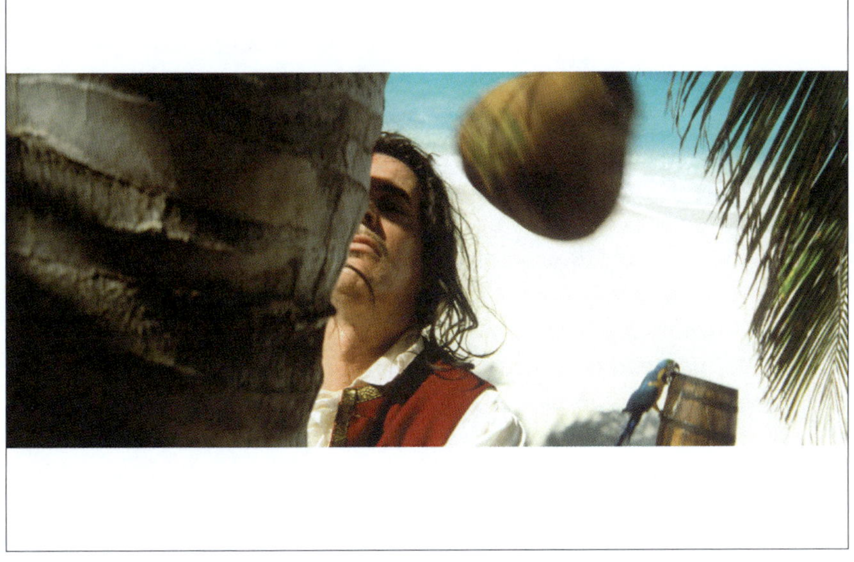

Figures 12a–12b. Rodney Graham, *Vexation Island*, 1997.

rendered, of the catastrophic events that conjoin states of consciousness (shots) according to at least one possible theory.'[36]

Boredom and repetition are convincingly embedded within this unbroken cycle of waking and unconsciousness. Graham's use of the word 'vexation' is a possible allusion to Erik Satie's famous piano work, *Vexations* (1893), a musical aphorism consisting of a thirteen crotchet enharmonic cycle to be repeated 840 times and to be prefaced by deep silence and inertia (the 'serious immobilities' indicated on the score). The piece was discovered by John Cage in 1949 and printed in facsimile as part of his *Contrepoints No. 6* that year.[37] Viewed as Satie's riposte to the ebb and flow dynamics of Wagner's melodic trajectories, *Vexations'* lack of harmonic or metric resolution and its construction based on the 'diabolical' tritone was associated with a musical curse or spell. With it, Satie condemns the performer to a constant reprise over many hours without reaching the satisfactory sense of an ending. Graham seems to echo this by instilling a languid monotony in the film with its infernal cycle of waking and falling. Each film shot follows the apparent logic of consciousness itself, rather than the pathway of narrative or film convention. This consciousness is not necessarily always that of the buccaneer. A bird's-eye shot might allude to a bird-centred consciousness and the parrot is as much a protagonist within the 'action' as the pirate.

By choosing a tropical island location, Graham points to ideas of the solitary castaway and the problem of human integration into nature, but the film goes deeper by exposing the misalignments between artist, performer, and the castaway trope in literature and film. Gravity is figured through the physical fall, as a falling into and out of consciousness. Falling objects are the inexplicable catalysts for each character's predicament: the falling hat in *City Self/Country Self* produces the pratfall of the bumpkin and the falling coconut in *Vexation Island* leads to the fall of the buccaneer. In *How I Became a Ramblin' Man*, the hoof fall of the horse in the third shot of the 9-minute loop is asynchronous with the film's non-diagetic sound, generated instead by two coconut halves knocked together to

<hr>

36   Rodney Graham, 'Siting Vexation Island', *Island Thought: an archipelagic journal published at irregular intervals* (Brussels: Yves Gevaert Verlag, 1997), 15–16.

37   Cage organized a performance of Satie's *Vexations* in September 1963 with a team of eleven pianists in a former vaudeville house.

complete the circuit of props and falls. For Graham, the looped events and falling bodies are like colliding particles. Rather than reflecting more indulgent ideas of subjectivity or identity, the characters of these films are the victims of other agencies, perhaps of blind nature itself, or their own subconscious drives. Thus, the suggestion of picaresque or action adventure is twisted into futile repetition with the buccaneer presented as a passive victim of other forces, whether psychological or natural. Artist Tacita Dean comments on how powerfully Graham renounces his own will within his practice:

> I don't know of any other artist who can quite do this, or who is willing to vacate their auteur self so completely, and then take up another role. It requires a level of surrender not normally undertaken by the artist, but one studied by the acting profession – to become the vehicle for someone else's expression albeit that that someone else was once, a short while before, themselves.[38]

Although Graham's interest in Freudian motifs of compulsion and repetition are at hand in these works, the focus on nature in the costume trilogy reflects shifts in the notion of material forces and non-human agencies. By casting himself into a role, Graham converts his body into a vehicle for external forces, emphasized in the slow-motion fall backwards towards the end of the loop. It is in falling, that the body most resembles an object with a shortcutting of will, autonomy, and agency. Our primary and very human defence is to withstand the vagaries of flux and chance, that experience of being no more than a particle awash in a sea of impersonal matter. Art, if considered as the act of controlling or ordering matter in the pursuit of free expression and a 'will to make', is somehow inverted here, at least at the level of content. Instead, self-alienation in the act of falling is a powerful strategy to undo our conscious volition, agency and will. In *Vexation Island*, causality is absent (what caused the bloody wound?), consciousness is hesitant or dimmed by a blow to the head with the buccaneer caught in a Sisyphean nightmare. Natural forces of wind, light, and gravity possess greater power than the comatose pirate. By his own admission, Graham's use of falling bodies is fundamental to his idea of the 'clinomatic instant',

---

38   Tacita Dean, 'Letter to the Aritst' in *Rodney Graham: Through the Forest*, ed. Friedrich Meschede (Ostfildern: Hatje Cantz Verlag, 2010), 57.

the chance swerve, wherein what appears as free will is only the fatal condition of entrapment.[39]

Recalling a terrifying childhood accident, Graham describes swinging
on a rope at the edge of a ravine in Harbour Chines and blacking out as
he fell into the ravine, his fall broken by a rotten log. Suddenly his body
was helplessly adrift, losing consciousness and balance at the same time.
According to the psychologist, William James, 'the primordial fact of
conscious personality' is 'the sentiment of volitional effort'.[40] Limit conditions, such as the fall, produce states of absolute contingency. But Graham
intuits the complex connections between the philosophical, physical, and
psychological parameters of falling. Both Graham and Ader present quite
melancholy understandings of the falling body as part of the fundamental declination of the whole world, revealing the pathos of our mistaken
belief in free will.

Post-human philosophy and the trend of new materialism create an
interesting context for understanding these repercussions of the body
as alienated from the self in the act of falling. Falling radically disrupts
the integrity of the body as subjectively willed action, and so helps us to
rethink the role of non-human forces as they impact on human subjectivity.
Questioning the hierarchy imposed by humanism and anthropocentrism,
Diana Coole and Samantha Frost signal a new mood that emphasizes the
agency lodged in other forms of matter:

> While new materialists' conceptualisation of materialisation is not anthropocentric,
> it does not even privilege human bodies. There is increasing agreement here that all
> bodies, including those of animals (and perhaps certain machines too), evince certain
> capacities for agency. As a consequence, the human species, and the qualities of self-
> reflection, self-awareness, and rationality traditionally used to distinguish it from

39   Steven Harris, 'Pataphysical Graham: A Consideration of the Pataphysical Dimension
      of the Artistic Practice of Rodney Graham', *Tate Papers 6* (Autumn, 2006) <http://
      www.tate.org.uk/research/publications/tate-papers/06/pataphysical-graham-consid-
      eration-of-pataphysical-dimension-of-artistic-practice-of-rodney-graham>, accessed
      20 August 2013.
40   William James 'The Feeling of Effort' in *Collected Essays and Reviews* (London:
      Longmans, Green & Co., 1920), 181.

the rest of nature, may now seem little more than contingent and provisional forms or processes within a broader evolutionary or cosmic productivity.[41]

This line of thinking could be extended to the question of how and why gravity becomes considered as an agent or medium within works of art from the 1960s onwards. It produces real effects, such as the drama of the fall, and works democratically on all objects with mass. In spite of its invisibility, gravity is the most stirring and ineluctable force and therefore exists as the most powerful counter agency to man. The chaos of the fall is the most credible visual symbol of the displacement of power away from man, the 'rational actor' to a situation wherein he loses any sort of leverage in the web of matter. Graham's use of costume indicates our own situation as actors, empirically locked in a cosmic drama and powerless to resist its stage machinery.

The Sorcerer's Apprentice

Throughout the examples given so far, there is a pervasive idea of how things or matter are animated or considered within a field of action. The oranges, shoes, coconuts, and dice described here are portable and cryptic objects that have been disengaged from causality and animated by play or lodged within a temporary *mise en scène*. Such objects have been transformed from use value into things that are opaque or resistant to function or meaning. The word 'prop' in its theatrical meaning severs the connection between objects and the notion of proprietorial ownership or commodity, rendering objects communal, portable, and flexible, able to be conjugated in different ways. Performance artist Stuart Sherman was particularly adept at re-imagining objects in his strange table-top performances held on street corners or at impromptu gatherings, producing

41   Diana Coole and Samantha Frost, eds, *New Materialisms: Ontology, Agency and Politics* (Durham: Duke University Press, 2010), 20.

failed magic tricks like the old British comedian conjurer, Tommy Cooper. Using manufactured bric-a-brac, Sherman reanimated disparate things into strange scenarios and unusual conjugations. Theatre prop cupboards provide glimpses of heterogeneous scenes, actions, spaces, and tableaux worked into different fantasies and illusions. In the case of the circus or music hall performer, props become the engine of work. Roland Barthes describes the aesthetics of music hall as the sublimation of work within an urban context wherein the performance is the precarious mastery of an action:

> A music-hall turn is almost always constituted by the confrontation of a gesture and a substance ... comics sculptors and their many-coloured clays, prestidigitators gobbling up paper, silk, and cigarettes, pickpockets and their lifted watches, wallets, etc. Now the gesture and its object are the natural raw materials of a value which has had access to the stage only in the music hall (and the circus), and this value is Work.[42]

More than simply rejecting the object as commodity or as a sign of work, these double acts of body and object erode the distinction between human being and thing to the recognition of a shared materiality of all things, of objectivized bodies and anthropomorphized objects. If objects such as skittles and balls are re-energized or given life in the juggling act, then this might start to be equated with approaches to object making in the post-war period. One way of drawing out the latent energy of found, manufactured objects is to place them in states of imbalance, as found in the work of British artist Richard Wentworth.

Wentworth's sculpture is characterized by his use of found domestic objects. His repertoire of dysfunctional buckets, slanting tables, and unstable ladders speaks of a world of workaday pragmatism from which our own repertoire of being takes shape. Gravity, often described as our sixth sense, is an experience so fundamental it slips out of our day-to-day consciousness and yet our language is shaped by the very 'up' and 'down' or 'heavy' and 'light' of our experiences. For Lakoff and Johnson, these are fundamental metaphors of being:

---

42   Roland Barthes, *The Eiffel Tower and Other Mythologies*, tr. Richard Howard (Berkeley and Los Angeles: University of California Press, 1997), 124.

> The concepts that govern our thought are not just matters of the intellect. They also govern our everyday functioning, down to the most mundane details. Our concepts structure what we perceive, how we get around in the world, and how we relate to other people. Our conceptual system thus plays a central role in defining our everyday realities.[43]

A hierarchy of needs imposes itself on the human psyche, transformed by the subject into habits of perception. Wentworth's art has been compared by Ian Jeffreys to the paintings of de Hooch with its delicate balance of ordinary domestic objects with the larger effects of light, space, and gravity.[44] If, as Hal Foster suggests, seventeenth-century Dutch *Pronkstilleben* displayed the treasures of an expanding commodity culture, then the Puritan simplicity of de Hooch's brooms, cupboards, bread, and linen were its spiritual counterweight. Wentworth instead catalogues objects of the late industrial age, invoking the 'thing power' of light manufacturing. His interest in gravity seems to follow the logic of Claes Oldenburg's assertion that gravity is the 'best form-giver'.[45] Wentworth's buckets, chairs, and tables are not quite subordinate to the rituals of everyday life and not integrated into a coherent vernacular of use. Instead, they elude the conventional taxonomy and order of domestic objects. *Twain* (1992) consists of asymmetrically hung weighing scales with a tilted, stainless steel cup attached to the weighing plate, suspended by a plastic cord. By subverting the possibility of measuring weight, it is as if the whole project of converting physical experience into a system of measurement is doomed to fail. By suspending or tilting objects, he inserts the 'shifty' diagonal into things. He interrupts the idea of the tool as subservient to human agency, instead revealing a vibratory force or materiality that emanates from the inanimate.

Wentworth's sculptures sometimes act like Beckett's double acts. Two buckets are soldered together to make a figure-eight formation, or two house bricks are conjoined by a section of ladder top. Sometimes the encounters

---

43   George Lakoff and Mark Johnson, *Metaphors We Live By* (Chicago: University of Chicago Press, 1980), 3.

44   Marina Warner, *Richard Wentworth* (London: Thames and Hudson, 1993), 14.

45   Richard Wentworth made this remark to me in passing as we were entering the auditorium at Tate Britain to listen to a Marina Warner lecture in 2002.

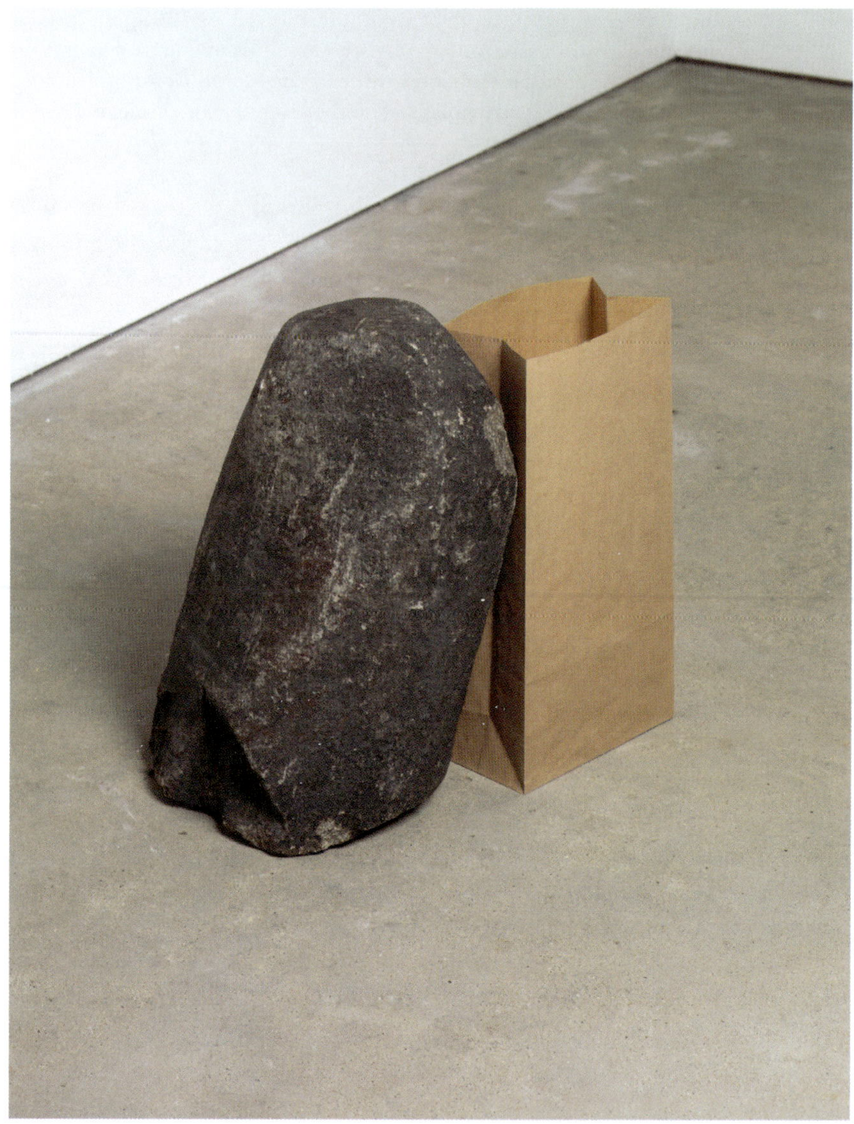

Figure 13. Richard Wentworth, *Domino*, 1984.

are odd bedfellows (a torch and a pillow in *Early Hours*, 1982); or evoke a politics of domination and submission (*Siege*, 1983–1984). In *Domino* (1984), a heavy rock is adjacent to a paper bag, the title inviting suggestions of a domino fall, even though the paper bag remains delicately resistant. Wentworth's objects do not serve us and, like disgruntled servants, start to present themselves in other incarnations or existences. As in the 'Sorcerer's Apprentice' scene in Disney's *Fantasia* (1940), brooms magically sprout arms and carry buckets of water, gaining a life autonomous of human control like that of an anthropomorphized mouse or Montaigne's knowing cat. If the artist is given to playing with objects from the junkyard, then the idea of tinkering might be said to correlate with this facet of artistic play.

The word 'tinkering' not only conveys the solitary activities of men or women in their sheds or studios, it also suggests mischievous behaviour or work done by idle hands. Wentworth's hybrid forms and assemblages evoke mysterious hands that undo the work of the factory or production line. By disrupting the assembly of hand-tool combinations, he reveals how objects slip out of their conceptual containers or taxonomies of use, acquiring other guises. Animating inanimate matter through modalities of falling, tipping, or balancing somehow distracts us from thinking of the unbearable heaviness of our own bodies. Henri Bergson notes our ability to be captivated by qualities of lightness and vitality:

> When we see only gracefulness and suppleness in the living body, it is because we disregard in it the elements of weight, or resistance, and, in a word, of matter; we forget its materiality and think only of its vitality, a vitality which we regard as derived from the very principle of intellectual and moral life.[46]

If, for Bergson, the body is inert matter dumped down upon the living energy of will and soul, then the object brought to life is the possibility of the body's comic resurrection. There is magic and solace in the image of dead things that possess self-organizing power and autonomy.

Invoking sensations of imbalance is also a crucial feature of a series of works by Peter Fischli and David Weiss, created and photographed in 1984

46 Bergson, *Laughter*, 31.

called *Equilibrium: A Quiet Afternoon*. Again, simple household objects are used for these balancing acts with an array of chairs, bottles, shoes, kitchen implements, and food choreographed as acrobatic still-lifes. In *Natural Grace* (1985), a wine bottle is perched on an apple in an eggcup with a plate supported on its cork, counterbalanced by a fish slice and a ladle. An onion in a net is dangled off the ladle. The resulting colour photograph reveals a minor *tour de force*, likened by the artists to a circus trick. In an interview for *Frieze* magazine, Jorg Heiser describes these works as 'the most absurd, gravity-defying constellations rivalling Chinese circus acts'.[47] The subtitle for the series reads: 'Balance is most beautiful just before it collapses.'[48] With reference to Adorno's formulation of the *Kunststück*, this moment of tension is that quality of technical bravura where defying gravity is the moment of 'look – no hands!' Technical brilliance that looks like magic actually reveals a kind of nothingness. It is the type of effervescence delivered by operatic singers in passages of *coloratura*, where sound bubbles up but does not drive any underlying thematic or formal construction. It does not matter if the ladle is in fact a fork or a sausage, since only its weight and ability to balance is of concern, with equilibrium being the only deciding factor. These diagrams of balance do not concern the quality of the materials in any way or even the particular form but simply wrest the object combinations from a moment prior to their 'fatal disintegration'.[49]

The *Equilibrium* photographs reflect the technical prowess of the circus trick, an energy force field which itself is only a profound nothingness or spectacular waste of effort. David Weiss explained this idea of objects behaving like circus performers in relationship to the artists' larger evolution of the idea in *The Way Things Go* (1987): 'I've always found that astonishing anyway – the way people always laugh when the next thing falls over. Because for us it was more like a circus act, trained objects.'[50]

---

47   Jörg Heiser, 'The Odd Couple', *Frieze*, 102 (2006) <https://frieze.com/article/odd-couple>, accessed 24 January 2017.

48   Jeremy Millar, *Fischli and Weiss: The Way Things Go* (London: Afterall Books, 2007), 72–3.

49   Theodor Adorno, *Aesthetic Theory* (London: Continuum, 2011), 244.

50   Heiser, 'The Odd Couple'.

Located in a dark studio or garage space (specially selected by the artists), the 'action' seems to emerge from the darkness with a suspended black bag spinning over a tyre. As the bag lowers, it sets up a chain reaction of energy transfer lasting nearly half an hour. Filmed only with simple edits rather than with cinematic effects, the camera simply pans horizontally across the action. Tyres, balls, plastic bottles, watering cans, and candles are harnessed together in a choreographed collapse, these quotidian objects made into 'spirited, living beings'.[51] Jeremy Millar comments on how the film seems to imply a whole cosmology, starting with a black bag and concluding with a final dissolve into mist:

> What follows afterward, as the camera pulls back to reveal the spinning bag, is a process of increasing differentiation, of separation. Almost every story of creation depends on such differentiation, of the land from the sky, of light from dark, of day from night.[52]

The disarmingly simple mechanics of *The Way Things Go*, which hides the extraordinary behind-the-scenes work of the artists to keep the plates spinning, is a playful exposure of the law of physics and its immanence in our day-to-day lives. Slopes and inclines are often the provocation within this little drama and might be considered a more impassioned, local response to physics, whilst being consonant with a renewal of atomism in contemporary philosophy and science.

In contrast to the slick movement and flow of the Honda advert, which plagiarized *The Way Things Go*, the objects in Fischli and Weiss's film look as if they might fail at any point. There are hesitations and sections that run slow so that the flow becomes only tenuously complete. Ultimately, a rear-guard action is staged against rote automatism. Millar notes: 'There are moments when, instead of acting automatically and with immediacy, simply falling or rolling, the objects seem to hesitate, as if reflecting on what it is they are about to do.'[53] If there is an uncertainty principle at work in *The Way Things Go*, it is actually a challenge to the fixed idea of clockwork

51   Heiser, 'The Odd Couple'.
52   Millar, *The Way Things Go*, 21.
53   Millar, *The Way Things Go*, 75.

motion cranked by the hand of God. The apprentice has usurped the power of the sorcerer, unleashing an unstoppable chain reaction.

By using gravity, artists attribute vitality to matter, arresting the promise of transcendental aspiration but also revoking human agency or free will. There is a return to Gnostic doubt, marked by a revaluation and reinvigoration of matter with objects brimming with energy and autonomy. Reconnecting with the life of matter and things, artists simultaneously decentre human subjectivity by parodying their own creative role. Douglas, Graham, Wentworth, Fischli and Weiss, and Wood and Harrison create small experiments or tableaux wherein the fall becomes the engine of creativity, energy, and play, and their own agency as artist-creators is taken apart through the techniques of *deus ex machina* or through a kind of artisanal modesty invested in throwaway objects. Hidden cuts in film, techniques of looping, or simply of the flashbulb or camera, are echoed at the level of content and refer back to duplication's dark agency. Agency is diverted onto inanimate objects, or distributed in the co-dependency of the double act, estranged from solo subjectivity or creativity. Artists produce a taut but often humorous dialogue with the everyday mini drama or comedy of having a body and of being placed in relation to the world of things without coherence or rational order. Gravity serves as a potent counter agent to man's hubris. Slapstick and pratfall indicate earth as a cosmic prison that nails us down. These artists create private cosmogonies and mediations, conducted in garages, sheds, and workshops in order to mediate and articulate that slipstream from the occult heart of matter to the pragmatics of daily life lived in continual declination.

CHAPTER 3

Heavy Stuff

In this chapter, I investigate how contemporary artists have interpreted a cultural imaginary of weight in shifting configurations of work-as-art or art-as-work. Weight is a profound condition of being. Our physical repertoire stems from the gravitational field in the way it forges the 'up' and 'down' or 'heavy' and 'light' of language and experience. Gravity shapes how the body connects with ground and how it confronts heavy lifting and the exertions of manual work. Represented as part of a larger machine, the factory body is a crucial link between mass, gravity and mechanical function. However, there is a corrective to the work of the manual labourer found in forms of leisure with cultural affects of lightness evident in new media experiences, such as cinema. It is as if the mass of an object, its inertia and resistance to lift and movement in the factory is evaporated under the widespread impact of mass entertainment.

After 1960, artists seem to assume identification with the manual worker making art out of the scrapyard, steel factory or shipyard. Richard Serra's process art and the minimalism of Donald Judd and Carl Andre appropriate the obtuse neutrality of factory production or steel mill production. Even the physicality of painting might be said to assume a work-like quality. Leo Steinberg argues that to the American artist, 'art' was guiltily connected to ideas of artfulness or artifice. Instead, Steinberg notes the invocation of honest, heavy work within American art, 'work' being somehow less tainted by association with foppish European high culture.[1] By connecting to the role of worker, artists, particularly from the 1960s onwards, entered into a complex dialogue or relationship with work as manifested in a broad spirit of enterprise and in identification with the

---

1   Leo Steinberg, *Other Criteria in Twentieth-Century Art* (Oxford: Oxford University Press, 1976), 60.

manual labourer's heroic physicality. As Raymond Williams argues, 'there could thus be at least a negative identification between the exploited worker and the exploited artist.'[2] For Williams, there is certainly a defensive gambit that speaks of how artists might try to refute those market forces that overwhelm the work of art. However, as the artist has encroached upon the factory, is this negotiation increasingly tinged with social embarrassment around issues of alienation with the creatively fulfilled artist usurping the materials and machines of the alienated worker? In art discourse, the word, 'fabrication' has become significant: it speaks of how large studios take on the function of the factory; it tells of how art making is increasingly delegated to technicians and in some cases of how art takes on the look of manufacture. As a result, 'fabrication' also indicates the erasure of the artist's authentic hand in the making process so that it is possible just to sign a configuration of bricks and steel after their anonymous manufacture.

Although writing just prior to the Industrial Revolution, Adam Smith's use of the term 'hidden hands' defended the free market's regulation of supply, demand and income; however, in a common form of myth, these 'invisible hands' of free market capitalism whisper something of a metaphysical order as if the market behaved according to some natural, hidden pattern, or providential plan. Somehow, the adoption of Smith's little phrase neatly occludes the actual exertions of real hands at work in countless factory spaces. In this context, gravity functions as a ground bass to labour in the correlation between a 'mechanics of force' and the worker's physical function in the factory: to lift weight and displace force. In calibrating workers' bodies, physics was the tool used to measure a standard or average ability to pull or lift weight. As developments took place in the measurement of work, of ballistics, of engineering, and manufacture, these were matched by the accretion of capital or expansion of territory. Gilles Deleuze and Félix Guattari summarize this relationship:

> There was a profound link between physics and sociology: society furnished an economic standard of measure for work, and physics, a 'mechanical currency' for it. The wage regime had as its correlate a mechanics of force. Physics had never been more social, for in both cases it was a case of defining the constant mean

2   Raymond Williams, *Politics of Modernism: Against the New Conformists* (London: Verso, 2007), 53.

value of a force of lift and pull exerted in the most uniform way possible by a standard-man.[3]

This tension between physics and wage regime is made explicit in the opening scene of Karel Reisz's *Saturday Night, Sunday Morning* (1960), when Nottingham machine operator, Arthur Seaton (Albert Finney) has just completed the lathe of 1,000 parts, the standard set for a day's work. He counts the parts precisely and makes clear the correlation between his work and his weekly wage of £14 3s 2d. Shot in grainy black and white, the sequence draws on documentary tradition in its unvarnished look at factory life with its deafening noise and endless machines. An early montage sequence includes close-ups of Seaton's hands, as he operates the machinery and then cleans his hands afterwards. This reference to hands is echoed soon afterwards at his home, as he tells his father the news that an operator injured his hand in the machinery. Seaton is part of a trope of the 'angry young man' in this period of British filmmaking with factory labour presented as an activity or an exertion that produces monotony and futility for the worker.

In the examples discussed here, gravity contributes powerfully to the appearance and disappearance of work or of a work force. The films of Salla Tykkä, Harun Farocki and Steve McQueen, as well as works by Richard Serra, Richard Wilson, Chris Burden, Theaster Gates and the controversial projects of Santiago Sierra unlock the spaces of labour as bodily submission or resistance to gravity. In these examples, gravity is tied to questions of work, war, and the new leisure industries, as contemporary artists have inhabited the sometimes culturally opaque regions of factory and industrial site.

Heavy Industry

In my calibration of weight and mass in relation to gravity, I need to backtrack a little and consider how the nineteenth century arises in the collective imagination as a world of heavy stuff. This was certainly a century in which

3 Gilles Deleuze and Félix Guattari, *A Thousand Plateaus: Capitalism & Schizophrenia*, tr. Brian Massumi (London: The Athlone Press, 1999), 490.

the forces of war and industry were significant for the cultural formations that followed. From a political point of view, the apparent inviolate power and permanence of vast engineering structures and machines is fixed in a position that reflects industrial capitalism. A famous image that exemplifies the nineteenth century's aggregation of weight, scale and force under gravity is Robert Howlett's famous photograph of the great Victorian engineer, Isambard Kingdom Brunel. Taken just a couple of years before his death, Brunel stands in front of the enormous launch chains of the *Great Eastern* ship at Millwall (1857). The photograph conveys a feeling of sublime power-in-mass, controlled and ordered by the great engineer. With a gross tonnage of 18,915, the Great Eastern was the largest ship ever built at the time of its 1858 launch, following the pattern set for Victorian 'leviathanism' in terms of bridge, hotel, railway, and ship construction. W. G. Sebald's melancholy vision of Europe is often focused upon locations tied to the industrial flow of goods and commodities, as well as the transit of people. His narrators inhabit Victorian hotels, railway stations, and ports, palimpsests of a great industrial past. The size and scale of industrial power is encapsulated in Sebald's description of the history of the Manchester Ship Canal in the final story of *The Emigrants*, based on the narrator's encounter with the painter, Max Ferber:

> The Manchester Ship Canal, Ferber told me, was begun in 1887 and completed in 1894. The work was mainly done by a continuously reinforced army of Irish navvies, who shifted some sixty million cubic metres of earth in that period and built the gigantic locks that would make it possible to raise or lower ocean-going steamers up to 150 metres long by five or six metres. Manchester was the industrial Jerusalem, said Ferber, its entrepreneurial spirit and progressive vigour the envy of the world, and the completion of the immense canal project had made it the largest inland port on earth.[4]

Sebald's narrator moves on to describe the decline of the docks from a glorious past to a motionless and silent canal, as he gazes back at the city 'sinking into the twilight'.[5] Industrial ports such as Gravesend and Manchester's Trafford Park bear the imprints of their colonial and industrial history, freighted with accumulated meanings of loss, profit, immigration,

---

4   W. G. Sebald, *The Emigrants*, tr. Michael Hulse (London: Vintage, 2002), 165–6.
5   Sebald, *The Emigrants*, 166.

emigration, and a 'dream-world of farewells'.[6] Such images of heaviness and stagnation relate to the writer's historical concerns, particularly evident in his indebtedness to Walter Benjamin. Critical examinations of late capitalism and the engines of heavy industry are inevitably bound up with the historical effects of the industrial ruin and junkyard.

Often the processes of heavy industry and labour conditions are not transparent or visible. Such blind spots are endemic to the forces of capitalism and globalization, obscuring the critical juncture between Marxian processes of alienation and reification. The thing must stand in place of the social and material relations that formed it, whilst the heavy lifting necessary to its fabrication and flow within a capitalist economy is screened through processes of exchange and commodity relations. Contemporary artists particularize these issues with the exertions of workers exposed, most clearly demonstrated in the projects of Santiago Sierra; or in the migration from the studio to practice within the sites of factory or industry, as in the work of Richard Serra.

In her analysis of the politics of participation, Claire Bishop has recently re-examined the relationships between industry and art in *Artificial Hells* (2012), assessing the ramifications of the APG (Artist Placement Group), active from the late 1960s through the 1970s at a moment when the coal and steel industries were still active in British society. In contrast to those artists in America, such as Judd, Andre and Serra, who worked with industrial fabrication from the early 1960s onwards, the APG was concerned with more finely calibrated interventions into industry, setting up a dialogue with the workers or management, rather than explicitly taking over the tools and machines of the workplace for their own art production. The idea for the APG came to Barbari Steveni, whilst she was out scavenging scrap and junk materials for artists, Daniel Spoerri and Robert Filliou on the Slough Trading Estate.[7] She recognized the potential advantage for artists to work in the factories rather than using the materials discarded outside.[8] Founded

6   Graeme Gilloch, *Myth & Metropolis: Walter Benjamin and the City* (Cambridge: Polity Press, 1997), 130.

7   Claire Bishop, *Artificial Hells: Participatory Art and the Politics of Spectatorship* (London: Verso, 2012), 164.

8   Bishop, *Artificial Hells*, 164.

in 1966 by Steveni and her partner, John Latham, the APG was conceived rather ambiguously as a dialogue between artist and industry through the placement of artists within large British corporations. Companies were not to expect any clear material outcomes but were instructed to remain open to the creative input of the artist immersed in their business. APG placements included Latham himself, who worked at the British Coal Board, Ian Breakwell placed at British Rail, Stuart Brisley at the Hille Furniture Company, and Garth Evans who went to work at Port Talbot for British Steel. APG's slogan was 'the context is half the work'. The outcomes were varied and piecemeal, though broadly speaking the aims were to instil a humanizing effect on the tough world of industry.[9] Nevertheless, criticism of the APG was targeted at the too close identification with corporate culture, characterized as a general failure to de-alienate the workers.[10]

Steel Works

As a young man from a working-class background, Richard Serra's early experience working in a steel mill reinforced an approach that involved a shift from traditional studio practice into one involving a process of negotiation with industry: 'Steel mills, shipyards and fabrication plants have become my on the road extended studios.'[11] Serra's massive sculptures are widely known and his interest in mass and gravity is well documented. One of his best-known works is *One-Ton Prop (House of Cards)* of 1969 (refabricated in 1986), constructed in a tough lead alloy, a material Serra

9   Bishop, *Artificial Hells*, 166.
10   It is worth mentioning the group, 'Experiments in Art and Technology' (EAT) based in America in the 1960s and 1970s, which offer a parallel to the APG but with a stronger focus on making work supported by techniques in industry, engineering and technology.
11   Richard Serra in Armin Zweite, 'Evidence and Experience of Self', ed. Ernst-Gerhard Güse, *Richard Serra* (New York: Rizzoli International Publications, 1988), 10.

had encountered whilst working in steel mills and shipyards as a young man. This freestanding sculpture invokes the tectonics of architecture as it is propped just through the interaction of its parts, producing a disquieting uncertainty for the audience in a sensation of imminent collapse. This work is somehow suggestive of Heinrich von Kleist's comment in a letter to Wilhelmine von Zenge of 1800, when he noted the uncanny ability of an arch to remain erect: 'Why, I asked myself, does this arch not collapse, since after all it has no support? It remains standing, I answered because all the stones tend to collapse at the same time ...'[12] At this point in his artistic development, Serra was reacting to the word, 'prop' in his *Verb List* (1967–1968) that, for him, brokered an interesting deal with process and materials. In the modality of propping, he rejects the welded joint, preferring to expose the tectonics of his sculpture. Welding elements or structure together belongs to the rationalizing and regulating tendencies of architectural form, but as he contends, 'any kind of joint – as necessary as it might be for functional reasons – is always a kind of ornament.'[13] Hal Foster argues that by emphasizing the tectonic in his work, Serra critiques the loss of the tectonic in architecture (particularly in relation to the surface qualities of post-modern architecture), just as he 'reasserts the rights of the body against the abstract objectivity and panoptical master of architectural concept and design.'[14] By testing sculpture to the point of collapse, Serra's works infringe the audience's sense of balance and uprightness. Instead of the body assaulting architecture's inert rectitude, as explored in Chapter 1, here the slanting and propping of heavy slabs assail the body's uprightness.

However, Serra's interest in the industrial processes of steelmaking and in the working conditions of steel workers is perhaps of more relevance here. In 1979, whilst working in Germany, Serra made a black and white film with Clara Weyergraf (*Steel Mill/Stahlwerk*, 1979, 25′6″) about the

12   Kleist's reading of the collapsing stones of the arch is explored by Jane Madsen, 'Kleist and the Space of Collapse', PhD thesis, UCL Bartlett School of Architecture, London, 2016, 10.

13   Serra in Hal Foster, 'The Un/making of Sculpture', *Richard Serra*, October Files (Cambridge, MA: The MIT Press, 2000), 185.

14   Foster, *Richard Serra*, 184.

social context of a steel mill in the Ruhr valley, where he was to carry out a commission for a site in Berlin. The mill (then taken over by *Thyssen*) was situated in Hattingen, one of many steel villages in the Ruhr valley producing metal for wheels, turbines, and nuclear reactors.[15] On an initial visit to the factory, he was particularly struck by the mill's forge, a colossal structure, almost 70 feet in height. Serra was encouraged to make a work at the mill by a gallery owner, which was eventually installed close to the train depot in nearby Bochum. The location was a satisfying choice, given its proximity to the place of its manufacture and its visibility to the steel workers, who used the train to get to work.

Serra became conscious of how the 5,000 workers employed in the rolling mills, scrap mill, and tool-and-die shops at Hattingen were never aware of the final applications of their steel products. He noticed the disparity between his complete identification in making his art within the steel factory and the alienation of the German steel workers at the mill. Serra's commitment to process rather than end product is well documented, but in this project he was made aware of how the fabrication of his work in the steel mill ultimately belonged to the domain of the fetishized artwork. This connection between industry and art was more like the relationship envisaged in William Morris's 1884 description of 'A Factory As It Might Be' in which 'there would no work that would turn men into mere machines' and in which beauty and meaningful labour would drive out ugliness and exploitation.[16] Nearly one hundred years on from Morris's manifesto, Serra was taken aback by the unhappiness of the workers, as revealed in a series of interviews, and was made to confront the myth of the 'happy, heroic German worker'.[17]

The film emerges as part documentary, part agitprop, shot in black and white, drawing on Serra's previous experiences of making process film.

---

15   'An Interview. Annette Michelson, Richard Serra and Clara Weyergraf' in *Richard Serra, Interviews 1970–1980*, ed. Clara Weyergraf (New York: The Hudson River Museum, 1980), 107.

16   William Morris, 'A Factory As It Might Be', *Justice*, April–May <http://www.marxists.org/archive/morris/works/1884/justice/12fact2.htm>, accessed 5 April 2017.

17   Weyergraf, *Richard Serra Interviews*, 108.

Using a constructivist style of filmmaking, he frames a critique of the mill's alienating modes of production. Images tracing the space of the factory are intercut with shots of workers at their machines, absorbed into a process that chains muscle to machine. *Steel Mill/Stahlwerk* reveals the oppressive conditions of the factory with its terrible heat and noise. Dialogue between the workers is made impossible by the noise, so that they have to use signs and gestures to each other to communicate during the working day. Reception of the film locally was poor, based partly on the perception that Serra had constructed shots, which framed fragmented bodies harnessed to machine parts. It was thought that this only reinforced a sense of alienation, however, any other strategy might have served to present the factory conditions in too positive a light. For the artist himself, *Steel Mill/Stahlwerk* was mainly a moment of demystification of the powerful connection between worker and material.

This representational issue is a paradox since David Nye's reading of the technological sublime is one in which factory machines, their repetition and complexity, were sources of wonder with the worker demoted to being a mere machine part, often absent from photographs of factory interiors. Nye writes, 'visitors to the early factory often attempted to comprehend the scene before them mathematically. Counting the number of yards of cloth woven or the tons of raw cotton processed was one way to grasp the immensity of the factory ...'[18] In order to comprehend the scale of enterprise, it was necessary to deal in numbers. Serra's sculpture is characterized by its overwhelming weight and scale, made more vivid through the way he tilts or slants vast slabs of Cor-ten steel. This relates to his interest in heavy industry as a site of production; however contrary to Nye's concept of the technological sublime, Serra's use of scale and gravity is gathered towards an abstract rather than mathematical order, one that explores the relationship between the body and objects of huge scale in a poetics of mass. From *One-Ton Prop (House of Cards)* of 1969 to his more recent works, the escalation in scale and mass have continued to produce powerful feelings of dislocation or collapse in audiences.

18   David Nye, *American Technological Sublime* (Cambridge, MA: The MIT Press, 1994), 115.

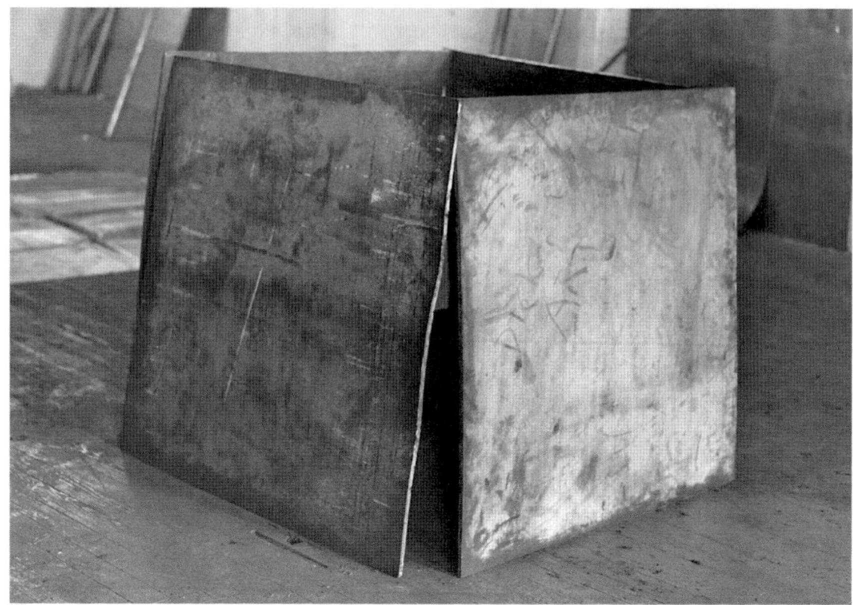

Figure 14. Richard Serra, *One-Ton Prop (House of Cards)*, 1969.

Telluric Gods

The use of industrial machines in making art has been a significant feature of post-studio practice since the 1960s: artists enlist tipper trucks, steam-rollers, fire engines, and newly invented machines in the actions of digging, scraping, and lifting. Poetic re-inventions of machines undo the work of industry, returning blind, autonomous technique back into magic. In *Earth and Reveries of Will*, Gaston Bachelard describes a psychotropism in the psychological feeling or anticipation of being crushed and its dialectical opposite, the will to withstand mass:

> The geographic dreamer – and such really do exist – steps in as Atlas to raise up the mountains, even at the risk of being taken for a braggart! Contemplating the terrain sympathetically, such dreamers throw themselves into the struggle of forces with all

of the conviction of a demiurge. To understand a mountain's mass truly, one has to dream of lifting it. The mountain brings its heroes to life.[19]

A collective imagination, fed by the overwhelming scale and mass of giant bridges, machines, and other forms of technological hubris in place of natural forces, produces a counter poetics of lifting and making light of weight. Duchamp humorously cited American engineering, bridges and skyscrapers as designs that validated his urinal as art in 1917. Leo Steinberg argues that the impulse to contain different parts in a whole painting surface was somehow paralleled by the development of the Detroit car with its fenders, wheels and mirrors corralled into a single design.[20] Performance artist Chris Burden's concept of a flying steamroller was realized in 1996 at MAK (*The Flying Steamroller*). He bought an ex-US Navy steamroller in 1991 at its scrap metal cost (it was a leftover from the Vietnam War when it was probably used for building roads or runways). Burden's main interest was in converting the 12-ton steamroller into a flying object through a system of counterweights and through the movement of the machine itself. The steamroller was to be driven in circles, just as the counter weight moved away from the central pivot, lifting the steamroller off the ground. As the momentum decreased, the counter balance would move back to the central point of leverage and the steamroller would land, re-assuming its full mass. This ironic negation of the engine's mass is a manifest symbol to challenge our fear of being ruled by machines.[21] In another work, *Beam Drop* (1984, re-staged twice in the 2000s), Burden created a sculpture at Art Park in Lewiston, New York by dropping sixty heavy steel beams weighing up to 2,000 lbs each into a pit of wet cement below. Each beam was lifted to a height of up to 120 feet by a crane before being dropped and, just after release, seemed to hover momentarily in the air, before plunging downwards. He explains the resulting sculpture, created through the interaction

19   Gaston Bachelard, *Earth and Reveries of Will: An Essay on the Imagination of Matter*, tr. Kenneth Haltman (Dallas: The Dallas Institute Publications, 2002), 277.

20   Steinberg, *Other Criteria*, 79.

21   Peter Noever, ed., *Chris Burden: Beyond the Limits* (Ostfildern: Cantz Verlag, 1996), 11.

of the weight of the beams and the chance falls produced by gravity, as an enlarged version of abstract expressionist brushwork with chance effects set against the certainties and rationalities of architecture. In *Beam Drop*, Burden's use of chance and gravity thwarts the orthogonal rationality of the steel beam. Many of his objects are based on re-contextualizing technological and industrial objects to create an understanding of humankind's general enthrallment to their power. He subverts conventional attitudes to machines, uncovering the rationalizing violence they do within a politicized field of vision. As Donald Kuspit confirms:

> Thus Burden cures us of our unconscious terror of the machine. At the same time, conceiving it as a toy suggests that we are in terror of it because it projects and represents our infantile aggression and destructiveness – our insatiable will to power, ostensibly in the service of self-preservation, in a social disguise. The machine is a sophisticated instrument in the service of a primitive instinct. ... We think we control our machines, but they secretly control us ...[22]

Within this negotiation between body, mind and machine stirs an imagination of weight that relates to the telluric gods of *Hercules, Samson, Sisyphus, Antaeus,* and *Atlas.* These figures of power are conceived of as positive symbols of human resistance. Gravity is a sedimentary condition of being so that weight is often mythically contemplated through narratives and images of resistance. Within the domain of art practice, such mythical figures are sometimes used to unlock dimensions of contemporary experience. Artists such as Burden recall fantasies of the resistant body in order to reveal a generalized amnesia about the body within a cultural trajectory that tends towards dematerialized affect. In *Samson* (1985), Burden installed a 100-ton jack together with gearbox and turnstile at the Henry Art Gallery, University of Washington. As visitors moved through the turnstile, the jack pushed two large horizontal beams against the load-bearing walls of the museum. Aimed at the ideological power of the museum, *Samson* could in theory trigger the literal collapse of the building, echoing and increasing the feeling of jeopardy induced by Serra's *One-Ton Prop*. Burden's structural

22   Donald Kuspit, 'Man for and against Machine' in Noever, ed., *Chris Burden*, 67.

installation can be seen to expose those mediating forces between physics and bodies, reflected in the structure of buildings and machines.

In 2012, Theaster Gates installed *Raising Goliath* at the White Cube, Bermondsey as part of his solo exhibition, *My Labor is My Protest*. For this exhibition, Gates lifted a red Ford fire engine in a symbolic gesture, counterbalancing the truck with a container of back issue magazines such as *Ebony* and *Jet*. *Raising Goliath* is conceived as a physical event that reflects a lifting-into or out-of-consciousness of the Civil Rights movement within the process of historical change. Gates's work is a complex mix of regeneration, political history, and labour history. Working from the south side of Chicago in a neighbourhood long associated with poverty and deprivation, the artist is keen to expose not only histories of slavery and exploitation, but also to revive the agency of the men and women living in this area.

In a BBC interview with Will Gompertz for the series, *Zeitgeisters*, Gates talks of the profound sense of humiliation and despair experienced by men from the almost exclusively black neighbourhood of Dorchester, when they were out of work and unable to support their families.[23] This might be compared to the miners interviewed in Jeremy Deller's *The Battle of Orgreave* (2001), who were faced with a crippling loss of livelihood during Thatcher's brutal suppression of the South Yorkshire mining industry in 1984. Gates's levitated Ford fire truck converts physical weight into symbolic value, revealing 'how important mythmaking is to how we believe in things'.[24] Within this discussion of how labour is mythified or occluded, Gates shows how these projects are a reinvestment in the role of the skilled hand. With a future world that remains 'tech invested', he believes: 'The more we can create a skilled hand, [the more this] will create new sectors of opportunity' in what he terms 'radical neighbourliness'.[25] This revaluation of the skilled hand recalls William Morris's reimagining of the hand's role in making beautiful things, wonderfully subverted by Jeremy Deller in his work for the British Pavilion at the Venice Biennale in 2013 when he enlisted

---

23   Theaster Gates, 'Zeitgeisters' [podcast] (21 July 2014), BBC Radio 4 <http://www. bbc.co.uk/programmes/b041469j>, accessed 8 August 2016.

24   Gates, 2014.

25   Gates, 2014.

artist Stuart Sam Hughes to paint an enormous mural of William Morris as an avenging and destructive colossus throwing Roman Abramovich's luxury yacht into the lagoon.[26]

Human labour is framed in the work of the Spanish activist-artist Santiago Sierra, who draws on the descriptions and language of work as part of his socio-critical brief. Sierra's performances centre on forms consonant with a loose industrial language of blocks, beams, and cubes, mirroring the abstract forms of minimalism. However, Sierra favours cheap and fragile materials or dull blocks of concrete, more reminiscent of crude construction or warehouse storage. Eschewing aesthetic associations with high-quality materials, Sierra focuses attention on the work and the workers involved in the performance. Weight-bearing is turned into the literalism of work. In 2001, Sierra organized a delegated performance at the Kunsthalle der Hypo-Kulturstiftung in Munich (*Raising of Six Benches*). He gave the instruction that six benches were to be raised for two hours a day for two weeks, with the variable heights set by the workers in accordance with their ability to lift the weights and in light of their own fatigue. At Deitch Projects, New York, in June, 2002, a group of three to four workers, found through local job centres, were paid $12 an hour to work as human caryatids, supporting *9 Forms of 100 x 100 x 600 cm* (constructed out of wood and asphalt). Again, Sierra rejects any possible connotations of an autonomous aesthetic, with the work's condition resting on 'rawness' and 'the sweat of hard work'.[27] Even the trappings of the work needed to manoeuvre the blocks into place are left with tables, scaffolding, pallets and packaging preserved in situ.[28] Sierra enlarges boredom, repetition, and the exertion of resisting gravity in these projects, exposing the real *mechanics of force* placed on the body within the capitalist wage system.

In *300 Tons*, created at *Kunsthaus Bregenz* in 2004, Sierra took weight-bearing to an extreme by placing 292 tons of concrete bricks on the top floor

26   Abramovich's guards had cordoned off the yacht, impeding the pedestrians walking to the Giardini at the previous Biennale.

27   Eckhard Schneider, ed., *Santiago Sierra, 300 Tons and Previous Works* (Cologne: Walther König, 2004), 42.

28   Schneider, *Santiago Sierra*, 43.

of the museum building. This huge weight was distributed on supports over the museum and would only allow entry for a maximum of 100 visitors at any one time (this additional average weight calculated at 8 tons). As in Chris Burden's installation, *Samson*, a turnstile was erected. As visitors entered *300 Tons*, the turnstile registered digital readings of the number of visitors and the projected difference between the permitted maximum of 300 tons and the current reading.[29] In these works, the threat of collapse, as posed by Serra's *One-Ton Prop*, is massively exaggerated. This was intended to make every visitor feel his or her own contributory weight in terms of any imminent collapse, making politics out of the audience's participation in a large cultural institution.

At ACE gallery in Los Angeles, Sierra organized a performance, in which workers were paid to move twenty-four blocks of concrete (*24 Blocks of concrete constantly moved during a day's work by paid workers*, 1999). Then at Kunsthalle, St Gallen in Switzerland, Sierra hired six Albanian refugees, without work permits, to move three cement cubes by hand for payment (*3 Cubes of 100 cm one each side moved 700 cm*, 2002). Severed from the goal of making something, the exploited workers are hooked into meaningless actions. As Anna Dezeuze writes: 'Sierra's dark work takes the form of a basic economic transaction, in which individuals are paid and exploited as objects following the logic of capitalism.'[30] By enlisting participants into a display of their labour, Sierra flushes work back into consciousness and makes its alienating effects all too vivid. This is of course not a new artistic phenomenon and in contrast to such macho lifting works, feminist art of the 1970s had revealed the unseen domestic labour of women.[31] In 1977, Mierle Laderman Ukeles became artist-in-residence at the New York Department of Sanitation. She made a series of 'work ballets' from 1984 to 2012, co-ordinating performances involving garbage trucks, barges, tons of waste, and sanitation workers. Harnessing a labour force in mythic fashion

---

29   Schneider, *Santiago Sierra*, 26.

30   Anna Dezeuze, *Almost Nothing: Observations on Precarious Practices in Contemporary Art* (Manchester: Manchester University Press, 2017), 242.

31   Kathy Battista, *Renegotiating the Body: Feminist Art in 1970s London* (London: I. B. Tauris, 2013).

is a critical feature of Francis Alÿs's *When Faith Moves Mountains* staged in Peru in 2002 with 500 student volunteers asked to move an entire sand dune with shovels. Alÿs's guiding mantra was 'maximum effort, minimal result' reversing the normal trajectory of industrial efficiency and somehow echoing Adorno's asymmetry between effort and result in the work of art and circus trick. As in Sierra's performances, labour becomes the goal in itself without trajectory or meaning; moreover, Sierra, Gates, Deller, and Alÿs complicate the question of agency through involving teams of men, women and communities in order to realize their projects.

Congo Line

In his short essay *On Lightness*, Italo Calvino notes the shift from heaviness to lightness within contemporary material culture: 'The second industrial revolution, unlike the first, does not present us with such crushing images as rolling mills and molten steel, but with "bits" in a flow of information traveling along circuits in the form of electronic impulses. The iron machines still exist, but they obey the orders of weightless bits.'[32] Lightness and heaviness are in constantly dynamic correlation with each other. Marshall McLuhan coined the term, 'surfing' to denote a rapid, multi-dimensional movement through information and also, as early as 1951, uses the image of the maelstrom (drawn from Edgar Allen Poe) to describe the electronic whirlpool of new media.[33] For Zygmunt Bauman, this move to lightness and speed, found in information, communication and technology is now the site-less position of contemporary power.

> It is now the smaller, the lighter, the more portable that signifies improvement and 'progress'. Travelling light, rather than holding tightly to things deemed attractive

---

32 Italo Calvino, *Six Memos for the Next Millennium*, tr. Patrick Creagh (London: Vintage, 1996), 8.

33 Marshall McLuhan, *The Mechanical Bride: Folklore of Industrial Man* (New York: Vanguard Press, 2002).

for their reliability and solidity – that is, for their heavy weight, substantiality and unyielding power of resistance – is now the asset of power.[34]

Such lightness, however, has a price that is much darker and weightier, since the miniature circuit boards of smartphones, laptops, and tablets rely on the hidden ingredient of rare earth metals such as coltan mined in appalling conditions in China and Africa. Simon Starling's *One Ton II* (2005) was included in an exhibition (*Simon Starling: Recent History*) at Tate St Ives in 2011.[35] The work consists of five platinum prints of an opencast platinum mine in South Africa. The central conceit of the work was that it required the mining of a ton of platinum ore to make just five prints. This extraordinary trajectory of heavy to light reveals the disproportionate amount of energy needed to produce just five images. Starling's *One Ton II* reveals the true cost of the West's desire for lightness and dematerialization through examining the processes and applications of metal mining.

In his 2007 film, *Gravesend*, artist and filmmaker, Steve McQueen exposes an unseen link between global capital's manufacture of 'light' media, represented by telecommunication devices such as mobile phones and laptops, and the raw metal extracted through 'heavy' industrial processes, exposing a highly specific relationship between the commodity-lite thing and heavy labour. He draws out the real human cost paid in the pursuit of a transcendent lightness in new electronic devices. McQueen's film is composed of a counterpoint of shot sequences: a sterile, high-tech British laboratory, in which technicians purify a metallic ore called coltan, using advanced precision tools; a mine in the Congo, where the metal ore is drilled out by workers working in appalling conditions; sunset shots of the industrial port of Gravesend with its chimneys and machinery shadowed against the setting sun and, towards the end, the juxtaposition of a coiling

34	Zygmunt Bauman, *Liquid Modernity: Living in an Age of Uncertainty* (Cambridge: Polity Press, 2000), 13.

35	'Simon Starling: Recent History' <http://www.tate.org.uk/whats-on/tate-st-ives/exhibition/simon-starling-recent-history>, accessed 30 January 2017.

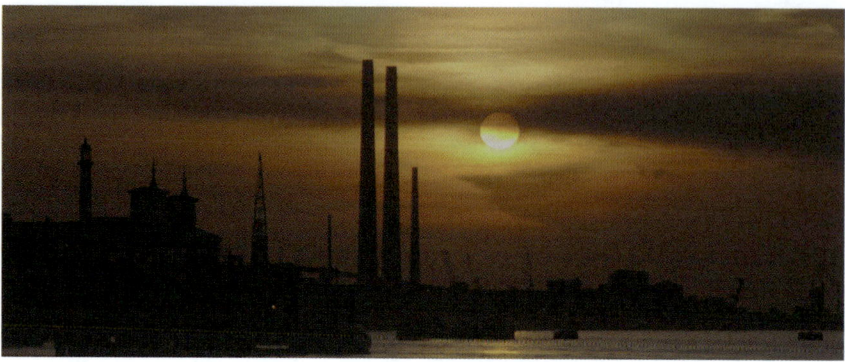

Figure 15. Steve McQueen, *Gravesend*, 2007.

black form and a white ground, suggestive of both the coil of fiber-optic cable and the sinewy flow of the Congo river.[36]

The industrial port at Gravesend holds resonances in terms of British colonial history, particularly with regard to Joseph Conrad's *Heart of Darkness* (1899). In the novel, the main character, Marlow is situated on the Thames estuary at Gravesend, while he narrates his experiences on the Congo River. McQueen's shots of Gravesend then segue into images of miners in the Congo, working in grim conditions to extract the mineral from the rocks. In his analysis of the film, T. J. Demos notes how McQueen's very particular approach avoids straightforward documentation to pursue far more ambiguous image-making, in which the Congolese miners are not clearly delineated or brought into clear focus, but are left as shadowy forms like vague abstractions. These qualities of image are contrasted with the hard-edged contours of the laboratory with its protocols of precision and sterility. Visual abstractions of these exploited figures might be thought to play into the same political position that keep the workers out of the visual regime of politics, but Demos notes another strategy at work:

36 T. J. Demos, 'Moving Images of Globalization', *Grey Room* 37 (2009), 6–29. doi/abs/10.1162/grey.2009.

One risk in this regard may be that the film's impressions merely reaffirm its figures' invisible status, reiterating their nonrepresentability in the register of the image. However, Gravesend's gambit is to draw out the very ambiguity of being so that life's separation from politics cannot disclose a simple ontological truth but rather must be viewed as political effect.[37]

This political point is reinforced by McQueen's mobilization of darkness as a trope, figured in the sunset, dark chimneys, shadowy forms, and mines, which sets up a circularity of association mirroring the real obfuscations created by a globalized economy. The image of Gravesend with its cranes, dock buildings and chimney stacks is reminiscent of W. G. Sebald's description of Manchester's industrial heartland as 'one solid mass of utter blackness'.[38] Similarly, the anonymous bodies and hands that, pitted against gravity, move and lift within the factory or mine are part of that cycle that turns everything into the abstraction of money. Images of work and industry remain remarkably opaque to understanding and there is a way in which perhaps the essential component of the human body within the factory remains strangely absent within visual fields of representation, just as the lived complexity of mass industrial working practices are difficult to assimilate within a shared understanding.

Ships and seafaring were crucial to the expansion of trade and governed by deadweight tonnage, displacement tonnage and load lines. Overcoming the ship's weight and inertia at its launch is an ordering of its gravity into lightness and movement, a moment that the artist Richard Serra acknowledges as a significant influence on his practice. Witnessing the launch of a tanker in Brooklyn in his youth, Serra remarks on how this shift between heaviness and lightness was a primal scene for his lifelong investigations into gravity and weighty materials. ('The ship went through a transformation from an enormous obdurate weight to a buoyant structure, free, afloat, and adrift. My awe and wonder at that moment remained.')[39] In his novel, *Rings of Saturn* (1995), Sebald invokes a meeting between the merchant

<hr>

37   Demos, 'Moving Images', 13.
38   Sebald, *The Emigrants*, 168.
39   Serra in Foster, *Richard Serra*, 193.

sailor-turned-writer Joseph Conrad and Roger Casement.[40] Casement and Conrad had shared certain views about the brutal inhumanity done to the Congolese during the 'Scramble for Africa', particularly the exploitation wrought by King Leopold II of Belgium. The Congo in McQueen's film *Gravesend* traces a return to the Thames, flushing its dark history back into present view.

Like McQueen and W. G. Sebald, Damián Ortega evokes Conrad's *Heart of Darkness* with the river journey indicating the flow of goods within the cycles of colonial exploitation. In *Traces of Gravity* at White Cube, Mason's Yard in 2012, Ortega created an installation of car tyres running through the gallery space to represent the Congo River with a line of salt spread across the surface, symbolic of the trafficking of cocaine.[41] The artist includes an overturned 9-metre long plastic copy of a Second World War German submarine with salt falling through the sculpture onto the floor to reference the recent illegal trafficking of cocaine in submarines running under the water between South America and Mexico. Flow, force, and gravity are symbolic markers of those unseen realms of raw material and commodity transit.

In her catalogue work, *W. G. Sebald* (2003), Tacita Dean discovered, in another strange historical return, that she was related to the judge who had condemned Roger Casement to death.[42] Just as Sebald engages seaports and ships as haunted spaces of an imperial past, Dean also explores the space of the sea as a dark reading of loss and failure. In turn, her film, *Disappearance at Sea* (1996) about the last voyage of Donald Crowhurst, was one of a series of works about the failed yachtsman. Another footnote uncovered by Dean was the fact that Bas Jan Ader had a copy of *The Strange Last Voyage of Donald Crowhurst* in his locker at Irvine, prior to his own disappearance at sea in 1975. In contrast to Hegel's myths of transcendence and historical progress, Dean, Sebald, and McQueen relate ports and ships to a ruined modernity and failed colonial past, symbolic of a

40   W. G. Sebald, *Rings of Saturn*, tr. Michael Hulse (London: Vintage Classics, 2002).
41   'Traces of Gravity' <http://whitecube.com/exhibitions/damin_ortega_traces_of_ gravity_masons_yard_2012/>, accessed 30 January 2017.
42   Tacita Dean, 'W. G. Sebald', *October*, 106 (2003), 122–36.

non-progressive history of eternal returns as invoked by Walter Benjamin. Such ruins are described in Todd Presner's appraisal of Sebald's *Rings of Saturn* as a melancholy litany of imperial loss: 'sunken ships, sailing logbooks, navigational instruments, model ships, crumbling piers, colonial relics and the remains of the shipbuilding industry'.[43]

The historical demise of shipbuilding and maritime industries in Great Britain has been explored in the work of artist Richard Wilson, most notably in *Slice of Reality* (2000). Having lived in areas of London bordering the River Thames most of his adult life, Wilson has developed an in-depth understanding of the river's history and the changes in its shipping industry.[44] Commissioned for the Millennium celebrations, *Slice of Reality* was a short section sliced out of an old 700-ton sand dredger called the *Arco Trent*. To make the work, Wilson had to take the ship to the River Tees, one of the few remaining shipbuilding areas. An engineering and welding firm then reduced the ship in length by 85 per cent, leaving a short vertical section of hull, bridge, and deck. One of the company directors saw the remaining sculpture as a metaphor for the merchant shipping industry in Britain: 'Sunk, shrunk and the heart torn from it.'[45] Re-using an old ship provided a critical counterpoint to the 'newness' of the twenty-first century being celebrated at the Millennium Dome on the banks of the river. Wilson cites Isambard Kingdom Brunel as one of his key influences, his infrastructural supports of engineering axiomatic in producing uniform environments of security and solidity. However, what he particularly admires is Brunel's ability to experiment with feats of engineering to the brink of disaster.

Wilson's site-specific sculptures introduce precarious imbalance with tilting or slanting floors, walls, and ceiling in order to challenge architecture's regulating constraints, as explored in Chapter 1. For the 2012 Olympics, the artist referenced the British crime caper, *The Italian Job* by re-creating the bus that dangled over a cliff edge in the final scene of the

43   Todd Samuel Presner, 'Hegel's Philosophy of World History via Sebald's Imaginary of Ruins' in Julia Hell and Andreas Schönle, eds, *Ruins of Modernity* (Durham: Duke University Press, 2010), 209.
44   Simon Morrissey, *Richard Wilson* (London: Tate Publishing, 2005), 68.
45   Morrissey, *Richard Wilson*, 71.

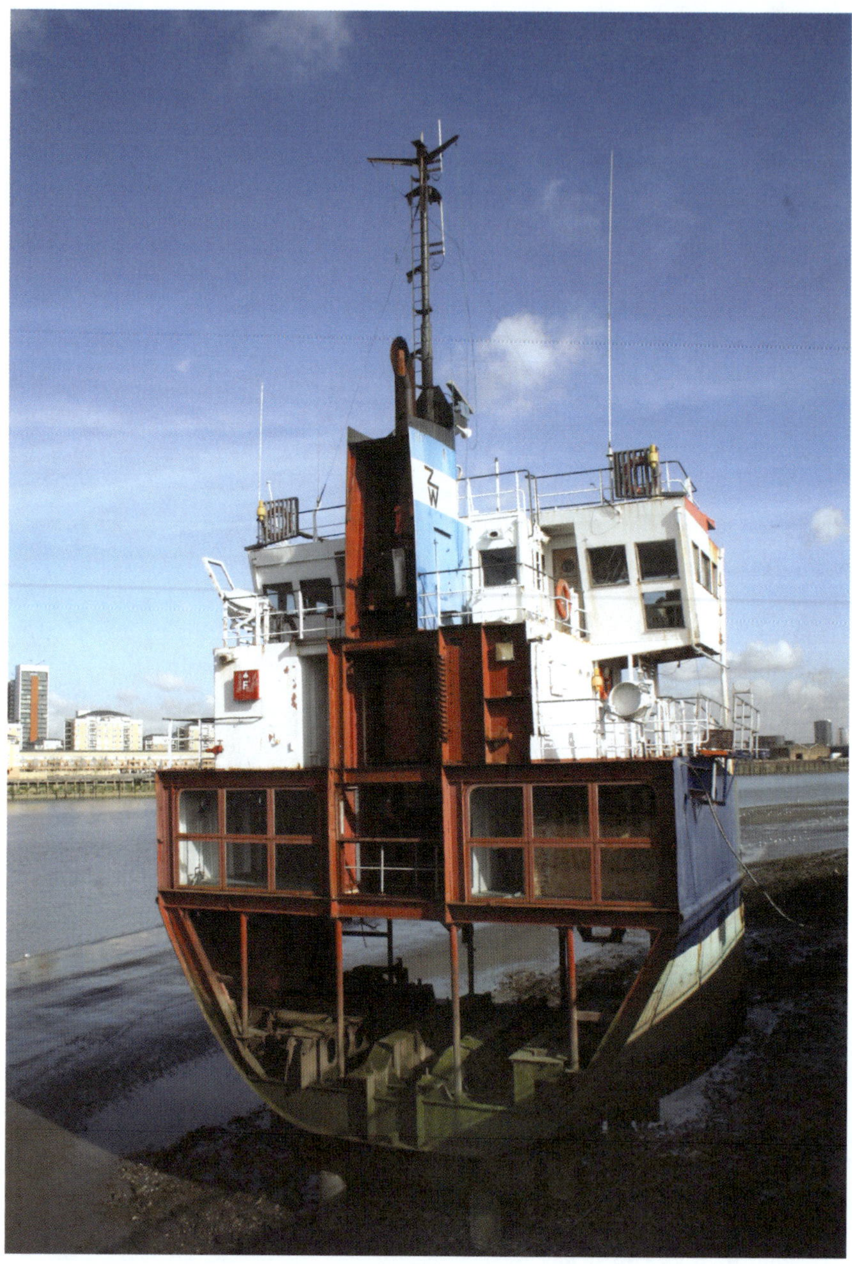

Figure 16. Richard Wilson, *Slice of Reality*, 2000.

film, taking its title from Michael Caine's last words, 'Hang on a minute lads, I've got a great idea'. The full-size replica was made to tip precariously over the edge of the De La Warr pavilion roof to reproduce a cinematic moment, which, according to Wilson constituted, 'a spectacle teetering on the edge of being and not being, and between stability and collapse'.[46] Periodically, the bus would slide yet further over the edge, heightening the tension of the piece for the audience, before tipping back up again.

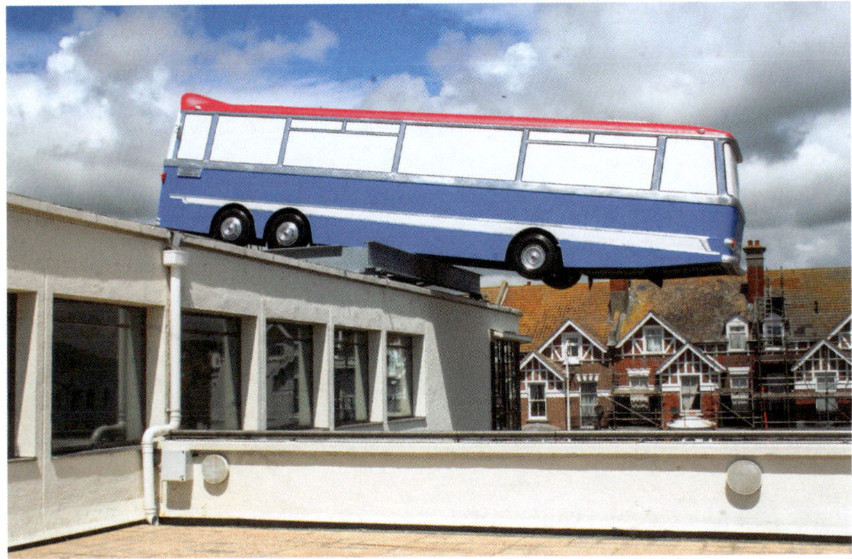

Figure 17. Richard Wilson, *Hang on a Minute Lads ... I've Got a Great Idea*, 2012.

Wilson's artistic response to gravity's challenge is complexly iterated since the decline of heavy industry has also been associated with working-class deprivation, particularly in the north of the country. Jeremy Deller's touring exhibition, *All that's solid melts into air* (2014), was based on how the ghostly afterimage of the industrial workplace resonates culturally now and what renegotiation of identity might be played out for the descendants or

46   Richard Wilson <https://www.dlwp.com/richard-wilsons-hang-on-a-minute-lads-ive-got-a-great-idea-goes-to-hong-kong/>, accessed 28 February 2017.

redundant workers of heavy industry. In terms of chronology, it was during the 1960s and 1970s, that the sharp decline in 'heavy industry' turned the steelworkers, shipbuilders, and miners into that melting condition of impermanence and obsolescence. There is, of course, a lag effect from the 1980s onwards, from an economy based on heavy industry to a service-based economy in relation to those re-valued connections between artist and engineer. Claire Bishop notes an overall tendency to model current labour patterns on the artist's project approach, which is mirrored in the general labour market as short-term contracts, outsourced service work-ers, and flexible portfolio working.[47] Just as the artist trading on personal creativity remains adaptable, so qualities of lightness and liquidity are central to the weightless flow of capital. The net effect within neoliberal economics is to mime the effects of freedom without the financial rewards of regular, permanent work. Still nourished by the spectacle and materials of old traditional industries, many artists remain operating in the spaces of heavy industry in attitudes of resistance.

Tour de Force

In the previous chapter, I argued that certain contemporary artists displaced agency within the artwork to an unknown or other force by using grav-ity's fundamental, yet mysterious operation. Behind the performance of *The Way Things Go*, Fischli and Weiss trained their objects into 'perform-ing' a set of tricks to produce a *tour de force*, on both a creative level and at the literal level of force. In terms of material, the Swiss duo chose objects consonant with old garages and workshops: used tyres, black plastic bags, ladders, redundant metal armatures, and old plastic cans. The stuff of light industry and car manufacture was dreamt into a redemptive afterlife by being turned into an abstract circus performance.

47   Bishop, *Artificial Hells*, 171.

As I reflected in Chapter 1, Adorno's affirmation of the circus trick in *Aesthetic Theory* centres on the defeat of gravity in the *tour de force* visible in the dialectic of performance. Performance is thus an arena of conflict, neither Hegel's 'ether' of ideal art, nor a fully pre-selected, mechanical reproduction. It is in the 'incompatible' demands placed on the performer to realize a tenuous relationship between the infinite potential of a work of art and its 'semblance' that the *tour de force* arises. As in Nietzsche's tightrope walker, performance is always predicated on near impossibility. It is perhaps the aspect of risk in the circus *tour de force* that attracted the historical avant-garde more generally with the performer continually threatened with collapse or the clown subject to the pratfall, the humour of which was agonistic to bourgeois ideology. Written during the 1960s but published posthumously in 1970, Adorno's *Aesthetic Theory* betrays greater acceptance of popular cultural forms, partly owing to his profound interest in the work of Samuel Beckett. He writes:

> The idea of art as a tour de force only appears fully in areas of artistic execution extrinsic to the culturally recognized concept of art; this may have founded the sympathy that once existed between avant-garde and music hall or variety shows, a convergence of extremes in opposition to a middling domain of art that satisfies audiences with inwardness and that by its culturedness betrays what art should do.[48]

As examined in Chapter 1, live performance in circus or vaudeville and music hall is the revelation of work and training aimed at producing a fragile bubble of something in performance that at the next turn collapses and disappears from view. This moment of suspense is critical to its value so that the sudden defeat of gravity is held out of time and historical flow and yet is powerfully subtended by work. Roland Barthes writes of this quality of suspension as an important element in historical painting, where the 'hero's caught gesture' interrupts time.[49] Acrobats, balancing acts, magicians are

---

48 Theodor Adorno, *Aesthetic Theory*, tr. Robert Hullot-Kentor (London: Continuum Impacts, 2004), 140.

49 Roland Barthes, *The Eiffel Tower and Other Mythologies*, tr. Richard Howard (Berkeley and Los Angeles: University of California Press, 1997), 123.

subsumed into the practice of their particular form of work. In his description of music hall, Barthes notes its rugged muscularity:

> What we have here is a way of making possible a contradictory state of human history: that the artist's gesture should set forth at one and the same time the crude musculature of an arduous labor, standing for the past, and the aerial smoothness of an easy action issuing from a magical heaven: the music hall is human work memorialized and sublimated ...[50]

One important consideration in terms of circus aesthetics is why an entertainment form so thoroughly entrenched within urban industrialized society was interesting to a modern avant-garde? It is imperative to keep Adorno's understanding of the circus *tour de force* in view. In Adorno's understanding, the work of art is continually overshadowed by its own 'fatal disintegration' or collapse.[51] This is its enigma. What attracted the historical avant-garde to circus was a complex mixture of an existential pathos associated with the lives of the performers (who had to conjure something out of nothing with only their bodies and the invisible force of gravity) and the vision of a sublimated danger contingent to the labouring body (held under the imperatives of capitalism and urban industrialization). *Why* indeed *waste all that effort?* Could it, therefore, be said that this vision of wasted effort in the *tour de force* harboured the labourer's perception of his own condition? Or, perhaps, circus performance was ultimately a vision of work without function or goal with bodies and objects transformed momentarily into images of mass-made-light and work-made-easy, sequestered from the unrelenting labour of the factory.

The trappings of circus were naturally emblematic of the routines and structures of administration, transportation, and commodification. Circus, as the dominant mass entertainment form enshrined the values, ideologies, and practices of a developing corporate order. During the high point of American circus, the railroad, the star system that cinema was to inherit, and new methods of billboard advertising reinforced the narrative of a total experience and the large circus mergers set up the tropes of monopoly

50   Barthes, *Eiffel Tower*, 124–5.
51   Adorno, *Aesthetic Theory*, 244.

capitalism and mass entertainment.[52] Strategies of how to move performers, equipment, and animals throughout the US were based on careful market analysis and a military style division of labour.[53] By globalizing entertainment experiences in this way, circus entrepreneurs laid down the tracks in which the new leisure form of cinema would move. However, in terms of its aesthetic, circus seemed to supply a free-floating experience freed from commodity relations, however illusory or temporary.

Workers Leaving the Factory

A number of recent exhibitions and films have probed the relationship between industry and the inheritor of circus, the cinema. *Against What? Against Whom?* was an exhibition of Marxist filmmaker, Harun Farocki's work at Raven Row, London (2009). The show included a multi-screen installation, which juxtaposed scenes of workers leaving the factory from the inception of cinema in 1895 with the Lumière Brothers' *Workers Leaving the Factory* through ten decades of the twentieth century. Interestingly, only four years separate the Lumière Brothers' first film, *Workers Leaving the Factory* (1895), and Georges Seurat's painting, *Circus* (1891). Circus and film are the forms of fantasy and spectacle subtended by repetitive labour, but which produce visions of bodies and labour-made-light. Seurat's painting emerges at a moment when the live circus was being increasingly displaced by cinema. If, as Slavoj Zizek generally argues, film is ideology in its purest form, then what are the full repercussions of an industrialized art that produces ungrounded bodies, both those within the cinematic spectacle and those who are suspended in the illusion, the cinema audience? Is there some faint displacement from the real ungrounded bodies of the circus acrobat, threatened by the abyss-fall and bodies magically ungrounded by

52 Janet M. Davis, *The Circus Age: Culture and Society under the American Big Top* (Chapel Hill: The University of North Carolina Press, 2002), 39.
53 Davis, *Circus Age*, 78.

the mechanics of the film camera? Certainly the ungrounded body in circus provided a rich trope of possibilities for cinema, leading from Chaplin to the complex kinetic types of burlesque comedy. This is then extended to the audience, whose bodies are equally ungrounded, caught up in cinema's spell of light and motion.

Farocki's *Workers Leaving the Factory in Eleven Decades* (2006) is an exploration of the cinematic apparatus, which is based on production line economics and the machine-like function of the camera. The work comprises twelve monitors placed in a line on the floor with each monitor showing clips from films from eleven different decades with most scenes set at factory gates. Sequences from films such as Fritz Lang's *Metropolis* (1926) and Chaplin's *Modern Times* (1936) reproduce the joint origin of cinematic production and labour found in the Lumière Brothers' first film, *Workers Leaving the Factory* (1895). For this first film of cinema history, the brothers filmed their own employees leaving work at their own photographic factory in Lyon. Farocki stages a 'cinematic return of the repressed' in that this installation foregrounds the general absence of the factory interior in film history.[54] Workers are filmed surging out of the factory gates in many of the examples as if released from the bondage of the assembly line across the threshold that marks out the work-leisure boundary. Like circus, cinema is a spectacle that supposedly compensates the worker and articulates his leisure-time consciousness. What Farocki exposes is how cinema disavows the work of the production line through its themes and affects, but how it is simultaneously based on a machine-like, industrial system of production.

Circus and cinema arose within an urban context and within a visual realm that supplied new perceptual possibilities, relying on the operations of modern machines to produce displays of levitation. Enraptured by the new and exciting kinetic possibilities offered by cinema, early filmmakers were drawn to the phantom ride, in which the camera was tripod-mounted on the front of a moving train so that audiences were seemingly transported through space along a railway track. At the 1902 St Louis Exhibition, Hales

54   Benedict Seymour, 'Eliminating Labour: Aesthetic Economy in Harun Farocki', *Mute Magazine* (14 April 2010) <http://www.metamute.org/editorial/articles/eliminating-labour-aesthetic-economy-harun-farocki>, accessed 4 April 2017.

Tours projected phantom ride footage inside an old stationary railway carriage to further enhance the illusion.[55] Early phantom rides were also created to spur demand for travel in the manner of a tourist commodity, such as the Edison phantom rides of Kicking Horse Canyon and White Pass, British Columbia (appropriated by Stan Douglas for his film montage, *Overture* of 1986). Simon Starling's *Phantom Ride*, commissioned for Tate Britain's Duveen Gallery in 2013, uses the experience of this early cinematic novelty as a method of collapsing time and space in relation to the real space history of the central gallery. Using a track-based camera, the artist traced the history of the Duveen Gallery from its bombing and destruction in 1940 to film and CGI images of the installations, sculptures and performances that have occupied the space since then. In 8 minutes of film, Starling concertinas history through a series of re-enacted works, all projected onto a 6-metre wide screen.

By harnessing machine movement, Starling creates a physical equivalent of cutting through time and space in the way that the cinematic medium first promised. Film before 1906 was fundamentally a 'cinema of attractions'. According to the film theorist, Tom Gunning, cinematic modes of visual experience are closely associated with the attractions of the fairground: 'The relation between films and the emergence of great amusement parks such as Coney Island at the turn of the century provides rich ground for rethinking the roots of early cinema.'[56] Early film took on the effects and novelties of fairground and circus in an unsettling of the body's stability. Fairground rides that use gravity conduct the audience's body into flights, falls, pendulum swings, and carousel rotations. Rollercoaster rides were in essence the relocation of railway traction and bridge construction to the amusement park. Audiences sought somatic pleasure in film, circus, and fairground, supported by technologies that move the body in powerful ways, as I examine more fully in the next chapter.

55 Mark Cousins, *The Story of Film* (London: Pavilion Books, 2011), 36.
56 Tom Gunning, 'The Cinema of Attraction[s]: Early Film, its Spectators and the Avant-Garde' in *The Cinema of Attractions Reloaded*, ed. Wanda Strauven (Amsterdam: Amsterdam University Press, 2006), 383.

Horsepower

The horse is an important point of mediation in our understanding of weight through the relationship between burden and bearer. The physical weight and pressure of the rider transmits commands by what Elias Canetti describes as the 'sting'.[57] There are strong sediments of thought that still cling to the horse's muscular force, evident in its connection to the idea of calibrating force (horsepower) and so measuring work. However, the horse was also the central act of modern circus associated with the military tradition of its founder, Philip Astley. It is useful to recover something of the nature of industrialized amusement shared by early film and circus. Jonathan Crary's fascinating research into Seurat's painting, *Circus* (1891) opens up a convincing cultural crossover from circus to the early cinema, since early cinematographs were sometimes placed within fairground or circus sites as novelties.[58] Crary has shown that the model for Seurat's alienated vision of the circus was most likely Emile Reynaud's Praxinoscope-Theatre in operation during the late 1880s.

Although Reynaud's artisanal contribution to dominant film history was his early form of celluloid technology, rather than photographic realism, his hand-coloured figures projected onto a screen point to the primal scene of that quasi-normative aspect of film: bodies moving without their feet. Seurat's horse is a spectral cut-out image, curiously detached from the ground and, according to Crary, 'the impossible frozen character of the image establishes it as a detachable or unbound component of a larger machinic synthesis or binding of images into the simulation of movement'.[59] In essence, Seurat's *Circus* can be understood as a frame from a larger continuum, an image taken from prototype cinema of the 1890s. If this painting is supposed to represent a suspended frame from an early film, then *Circus* is a striking premonition of the increasingly machine-like

---

57   Elias Canetti, *Crowds and Power*, tr. Carol Stewart (London: Phoenix, 2000), 367–8.
58   Jonathan Crary, *Suspensions of Perception: Attention, Spectacle, and Modern Culture* (Cambridge, MA: The MIT Press, 2000).
59   Crary, *Suspensions of Perception*, 278.

substrate of entertainment. The frozen quality of the image, with trick rider and horse hovering in air, records a moment of levitation, the effect of which was to be endlessly reproduced in cinema thereafter. It records Seurat's powerful insight into a critical juncture between the two dominant entertainment forms of circus and cinema, at a moment when circus was about to be displaced by film.

As Crary argues, given Seurat's insistence on scientific correctness, his luminous and ungrounded rocking horse is wilfully out of step with contemporaneous studies of horse locomotion, such as those conducted by Eadweard Muybridge at the Sacramento racetrack.[60] Seurat's spectral horse is far more convincingly indebted to a projected image from Reynaud's Praxinoscope together with its tumbling acrobat and crowd configuration. Seurat's choice of the locus of circus is therefore not reducible to a fantasy space (counter to the imperatives of factory life), but is as rudely mechanical as the space of industry. A more compelling argument is that circus entertainment offered the sort of kinetic promise conducive to new understandings of mechanized or cinematic vision. For Crary: 'The horse, which (despite the advent of the railroad) remained a primary nineteenth-century image of vehicular power and energetic motion, was part of that same early vocabulary of moving images and visual toys ...'[61] He sees the image of the horse and other circus figures caught up in numerous optical devices throughout the nineteenth century as embedded in a general drift towards industrialized vision.

In 2010, Finnish artist Salla Tykkä exhibited her short film *Airs Above the Ground* at the Hayward Gallery project space, based on the image and history of the Lipizzaner horse and its famous gravity-defying display. The film emerged at the same time as Frank Westerman's narrative history of the Lipizzaner, *Brother Mendel's Perfect Horse*.[62] These recent reprisals have examined the relationship between pedigree horse breeding and European history, warfare, and the *'Blut und Boden'* politics of fascism. Tykkä's film

---

60   Crary, *Suspensions of Perception*, 269.

61   Crary, *Suspensions of Perception*, 271

62   Frank Westerman, *Brother Mendel's Perfect Horse: Man and Beast in an Age of Human Warfare* (London: Vintage, 2013).

consists of over 7 minutes of film with intercut sequences of Lipizzaner foals running wild and a mature white Lipizzaner performing its 'airs above the ground', contrasting values of nature and nurture, freedom and discipline. From 1580, the Spanish Riding School in Vienna had developed this imperial breed in service to the Hapsburg dynasty and continued its tradition of balletic moves and extraordinary feats of levitation. According to Franz Westerman, 'airs above the ground' evolved out of battle tactics with the horse trained to kick the adversary, but its cultivation over centuries meant that this breed had a symbolism of far greater significance so that to touch a Lipizzaner was, in effect, to touch history.[63]

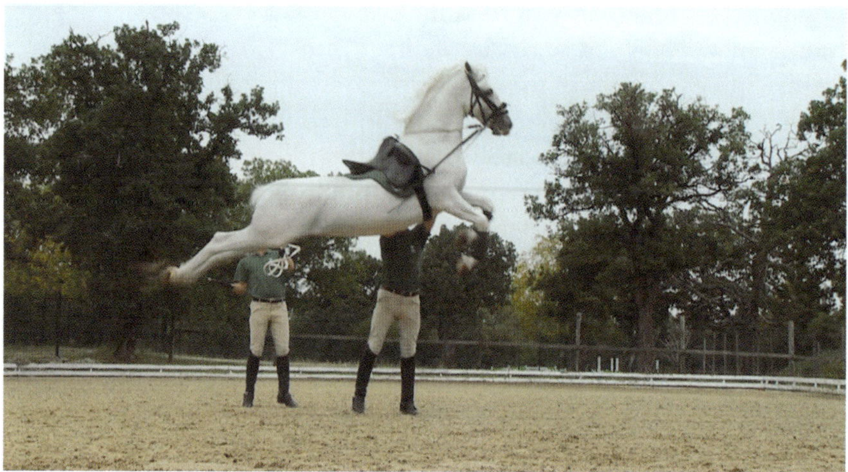

Figure 18. Salla Tykkä, *Airs Above the Ground*, 2010.

Accompanied by the sounds of its breathing and stamping hooves, the horse in Tykkä's film performs a *levade* in which the animal walks back a few paces and tucks its hind legs beneath its body in a controlled rearing action. As a culmination of years of intensive training, the Lippizaner is able to master the most difficult jump of all and extending itself from the

63   Westerman, *Brother Mendel's Perfect Horse*.

levade into the *capriole*, it is able to jump with all four legs off the ground, kicking its hindquarters out. This display of lightness is based on an extraordinary compression of energy and impulsion in the rump and hind legs, reinforced by careful breeding. Towards the climax of the film, the airs are shown in slow motion and accompanied by the *Agnus Dei* from Bach's B Minor Mass. The striking contrast between the horses running free within a natural landscape and the controlled training on a long rein within a ring reflects key tropes of the nineteenth-century imagination, the sublime image of unbridled energy and freedom encapsulated in Byron's poem, *Mazeppa* and hyperboles of power in Napoleon's requirement to appear 'calm upon a spirited horse'.[64] Tom Morton summarizes the meaning of Tykkä's film:

> This is a Pegasus-like fantasy of flight seeming come true, in which a member of a hereditary élite (indeed, a product of early-phase genetic engineering) leaves the fallen world behind, escaping the pull of gravity that afflicts other, earthbound creatures. The word 'air', in English usage, can mean an aristocratic affectation. The Lippizans' feats of dressage are precisely that – a grammar of gesture through which political power is performed and brokered.[65]

'Airs above the ground' reflect the horse's military associations and its connection to modern circus display in High School and Liberty acts. Tykkä's short film about the horse's transcendent display of power summarizes so many complex aspects of European history: of war, power, genetics (the horses become white at maturation), and work.

Rehearsal, repetition and rote mechanics, fundamentally important to performance, whether of horse, circus performer, or worker, expose the alienation at the heart of these displays. Activities of industry, war, sport, and entertainment are often consonant with each other; for example, with the foundation of body culture and physical movement, sport aped the 'work-out' of the factory. At the same time, the exertions of the worker have been euphemized or veneered by recourse to science, mechanization,

64   Martin Meisel, *Realizations: Narrative, Pictorial, and Theatrical Arts in Nineteenth-Century England* (Princeton, NJ: Princeton University Press, 1983), 216.

65   Tom Morton, 'Pale Kingdoms' in *Salla Tykkä: White Depths* (Bologna: Damiani, 2011).

and specialization in time and motion studies. The industrialization of vision, characterized by Muybridge's body motion studies, primitive cinema, and Etienne-Jules Marey's scientific measurement of the body and animal motion are joined at a moment when work was increasingly 'mythified' or sublimated within the new formations of the culture industry. Tykkä's film explores the arduous process of training the Lippizaner to perform its 'airs above the ground'. In dressage, the horse is forced into unnatural movements that require unerring patience and practice. Weightlessness and the beauty of momentary levitation are subtended by painful repetition. Nowhere is this clearer than in military and horseback spectacles. Interactions between body, ground, and gravity involve a wide array of cultural gestures and physical flourishes, which might reflect political or ethical statements of power or obeisance. Spectacles such as the Lippizaner's 'airs above the ground' convey an opaque symbolism that preserves the ghostly after-images of real function, such as work or war.

Repetition is taken to an extreme in the rote mechanics of circus performance. Circus takes the meaning of performance to a critical level. As Erwin Staus remarks: 'Performing means that we follow a given form, that we are able to conduct ourselves in accordance with a scheme set beforehand, that we can use our limbs like instruments.'[66] The singularity of certain bodily conditions or actions are emphasized in the flying, leaping, or balancing of the circus act. In *Stage Fright, Animals and Other Theatrical Problems*, Nicholas Ridout argues that performance art, with all its accidents and mistakes, is the repressed of the sleek (though sometimes faulty) mechanics of bourgeois theatre.[67] Noting how repetition in theatre is a nightly charade for the actor, he writes:

> [I]t is the knowledge that you are repeating something which is the problem with theatre. Not only is repetition a pressing problem of technique – the reconciliation of the necessity of repetition with the equally exacting necessity of apparent spontaneity … it also points to the disquieting possibility that the activities required of a theatrical performer are more like those of any other worker (the repetitive development of a

66   Erwin Staus, 'The Upright Posture', *Psychiatric Quarterly*, 26 (1952), 529–61 (543).
67   Nicholas Ridout, *Stage Fright, Animals and Other Theatrical Problems* (Cambridge: Cambridge University Press, 2006).

> skill and its daily exercise for wages), than they are like those of the new bourgeois idea of the artist, whose work is supposed to be spontaneous and free from the disciplines of a wage labour.[68]

Ridout reflects that the theatre actor is no more free from the constraints of repetitive labour than the factory labourer, but simply works within a domain that is, according to modern capitalism, the diametric opposite of work: leisure and entertainment. Max Horkheimer and Adorno's reading of repetition and the mechanical quality inherent in many forms of the culture industry shows how work-time consciousness and leisure-time consciousness must remain consistent, so that a film or cartoon moves in the 'worn grooves of association'.[69] With his re-valorization of the circus trick, Adorno notes how exertion becomes a key component in a labour sacrifice that produces a kind of nothing. This might also be the disappearing act that the performer rather than labourer creates in which work is transformed by sleight of hand or body. This is the trick-like or magical quality of the artwork. It is also a kind of nothing. Certainly, within circus, its nothingness is another demonstration or sign of a commodity-free zone. The performers set up this sign of autonomous freedom with their bodies, when in fact they are most particularly contained within the stranglehold of capitalism. When a circus performance contains an element of risk and danger, the need for rote training is more pronounced. Thus, the circus performer's reliance on repetition in fact stages the performer as alienated worker yet more vividly.

Simon Starling's *One Ton II* (2005), Richard Wilson's *Slice of Reality* (2000), Santiago Sierra's *300 Tons* (2004) and Chris Burden's *Samson* (1985) mobilize the effects of mass as a symbol or a medium to expose real histories and narratives of labour. Repetition and rote mechanics are circularities that, as the previous chapter explored, prove obstacles to progress. By short-circuiting myths of progress or glory, artists use the sites of vacant docks, and ruined factories to historicize the forces of industrialization and

---

68   Ridout, *Stage Fright*, 19–22.
69   Theodor Adorno and Max Horkheimer, *Dialectic of Enlightenment*, tr. John Cumming (London: Verso, 2016), 137.

capitalism in meaningful ways. Nevertheless, in the service of closer differentiation, it must be remembered that whereas traditional narratives of heavy industry have necessarily emphasized poor working conditions, long hours, and arduous labour, there has also been a reinvestment in a poetics of mass by artists, which rejects the inconstancy of current labour patterns, the shifty and fugitive nature of power, or the weightless flight of capital.

By investing in the structures or processes of heavy industry, artists such as Richard Serra are in fact holding onto the vestiges of work, place, and industry for their positioning within historical consciousness. As Foster states: 'Rather than fetishistic, then, his commitment to industrial structure can be seen as resistant – not only to the pervasive decay of the tectonic in sculpture and architecture, but also to its putative outmoding in a post-industrial order of digital design.'[70] Ruins, wasteland, and scrapyards are 'prime locations' for artists to unlock the collective rapture, amnesia, and trauma of power, will, and industrial might, spaces that might support true historical understanding. Just as gravity reveals itself as a powerful agent within a modern discourse of the tightrope, emblematic of art's balancing act between poles of reification and dematerialization, then weight is powerfully, if silently present within those cultural domains of work, sport, and the entertainment industries. Let us rephrase things – physics has perhaps never been so cultural, social and artistic.[71]

70   Foster, *Richard Serra*, 189.
71   Deleuze and Guattari, *A Thousand Plateaus*, 490.

Vertigo

A spaceless limbo on some spiral reels.[1]

If many material practices of art from the 1960s onwards were forged in the crucible of labour history, then what other artistic ideas flow from industrial processes? Robert Smithson took a different path in wresting meaning from the post-industrial landscape. He was drawn to disused industrial landscapes to make his art, such as oil sumps, obsolete mines, ruined factories, and old quarries; places latent with the drag and lift of mass, though lifeless as if abandoned by some hidden industrial demiurge. Smithson's implicit call to gravity is configured through vertigo, a mixed condition of lightness and falling, but gravity is also considered here within his larger concern for entropy. The artist's early interest in geology is expressed within his writings and practice, turned to a philosophical and artistic inflection of energy drain, loss of momentum, and deceleration, just as gravity slows time down. Entropy, for Smithson, is not just the collapse of a woodshed or decay of a site, but also an urban sprawl or the rust oxidizing on steel in contrast with the artistic fetish for steel and fabricated metal as found in the work of Judd or Caro amongst others.

In this chapter, I am concerned with the relationship of gravity to vertigo as mediated through Smithson's response to cinema within the scope of his multi-faceted *Spiral Jetty* project, and as connected to his short 1971 essay on cinema, 'A Cinematic Atopia'.[2] This article informs my approach to the

1   Robert Smithson, *Robert Smithson: The Collected Writings*, ed. Jack Flam (Berkeley, Los Angeles: University of California Press, 1996), 150.
2   Smithson, *Collected Writings*, 138–42.

complexities of Smithson's practice, explored primarily through his response to cinema and in connection to the film, *Spiral Jetty* (1971). Etymologically, the word, 'vertigo' derives from the Latin, *vertere*, 'to turn', a word that describes movement in a vortex. Vertigo is not just a dizzy turn at height, but is also connected to a more general feeling of disequilibrium. Vertigo can be described as a state that combines sensations of falling, lightness, and dizziness wherein the illusion of rotation and the ground heaving and shifting undermines the body's stability. Vertigo is not just a sensation related to the dramatic spectacles of height or the repressions of the tightrope walker discussed in Chapter 1, but is also manifest as a real experience that can be likened to the hallucination of motion felt by the cinema audience, as well as a more general metaphor of spatial dislocation and cultural fall. Movement, acceleration, and visual stimuli produce vertiginous experiences. Smithson expands vertigo's disorientation to include cinema's dislocations of space, reading the flickering screen and mechanical hum as symptomatic of the inertia and torpor it produces in the audience. Time, which is slowed by gravity, is, according to Smithson, also stopped in the movie house.[3]

It is first important to see how Smithson uses gravity. Selecting ruined industrial sites such as mines and quarries, Smithson was attracted to steepness and inclines in order to allow materials to comport with gravity, evidenced also in his allusion to a Brian Aldiss sci-fi chapter title, 'Down the Entropy Slope'.[4] One of his earliest 'flow' pieces is *Asphalt Rundown* (1969). In this work, a dump truck released a quantity of molten asphalt from the top of a gravel quarry near Rome, spilling downwards and outwards. In its final state, after the flow of asphalt stopped, it became an image of time arrested. In an interview with Dennis Wheeler, Smithson remarked: 'All the aspects of gravitational flow … all things move through gravity, you know, all the earth movements take place through some kind of gravity. And they're rather slow.'[5] Smithson makes gravity luminous in the processes he uses with materials such as glue, earth, rocks, and asphalt made to fall, drop, sink, and pour. The word 'gravity' sometimes occurs in his writing and drawings, but is

3   Smithson, *Collected Writings*, 10.
4   Smithson, *Collected Writings*, 215.
5   Smithson, *Collected Writings*, 216.

also implicit across a whole lexicon of collapse ('abyss', 'plunge', 'fall', 'dump', 'drag', and 'swoon', to name but a few). His complex iteration of entropy in his writings, films, photographs and earthworks is characterized by a relationship to gravity's pull. Gravity and time, co-ordinates in geological time, are the silent levellers that Smithson fixes into a visual experience, all ultimately aimed at deflating the hubris of human history or the Anthropocene. With customary irony, the artist also views gravity's devastating indifference as a modern affliction: 'Today, we are afflicted with an inversion of hyperbole – gravity. Rivers of Lead. Lakes of Asphalt. Heavy Water. Generalized mud. The Caretaker of Dullness – habit – lurks everywhere.'[6] Gravity's effect is seen as a deadening blanket that smothers hope, energy, and idealism. For Smithson, the dump truck or steep incline were ways of exposing gravity's indifferent tug, evident in his work, *Spiral Jetty* (1970). A quality of ironic negation is evident in Smithson's poured works, recalling Pollock's use of scale and force in that artist's dripping technique.

Located at Rozel Point on the Great Salt Lake, Smithson's 1,500-foot-long coiled earthwork is constructed from black basalt rocks, salt crystals, earth, and water, measuring approximately 15 feet wide. After acquiring land rights and locating a contractor with earth-moving equipment, Smithson managed the construction of *Spiral Jetty*. Two dump trucks, a tractor and a front-end loader were hired to shift 7,000 tons of rock and earth onto the site. Situated close to some defunct oilrigs and an old pier, the site attracted Smithson because of its stark appearance and what he described as a 'succession of man-made systems mired in abandoned hopes.'[7] The earth or ground exerts a drag that ultimately thwarts the ideal project. The ground is a shared condition that leads only to entropy and collapse. Smithson's selected sites, such as swamp, marsh and oil sump magnify the earth's tug. Swamp (*Swamp*, 1971), salt marsh, 'seeps of heavy black oil' and 'a quagmire of sticky gumbo mud' confound our false faith in stability, permanence, and idealism so that 'logical purity finds itself in a bog'.[8] Mud and swamp drag us down, amplifying or analogising gravity's pull, but also what runs

6   Smithson, *Collected Writings*, 127.
7   Smithson, *Collected Writings*, 146.
8   Smithson, *Collected Writings*, 146.

aground in this sucking or squeezing action of the earth is any ideal of form, structure or logic. Everything is ultimately reduced to undifferentiated matter. This statement resonates with Smithson's own founding idea for *Spiral Jetty*: 'From that gyrating space emerged the possibility of the Spiral Jetty. No ideas, no concepts, no systems, no structures, no abstractions could hold themselves together in the actuality of that evidence.'[9] *Spiral Jetty* is not confined to the earthwork of the same name, but comprises complex iterations across a variety of material forms: drawings, an essay, a film that includes a script from the essay, photographic documentation, mud, salt, crystals, rocks, and water. This radical distribution of elements, related also to his key bifurcation between site and non-site, is a meltdown of futile category.[10] Smithson equated this radical disruption of category and medium with a quasi-filmic perception. As Peter Osborne confirms: 'Smithson's imagination of the experience of the site of Spiral Jetty, it turns out, is actually the constructed result of his own film.'[11]

Light and speed seem fundamentally at odds with Smithson's fascination for slow geological time. That said, celluloid did not escape his reading of entropy and his practice was so often defined in opposition to the fatal seductions of cinema. As far as Smithson was concerned, film provides only 'false immortality' since the celluloid itself will still 'crumble or get lost and enter the state of irreversibility'.[12] Something darker was also hidden within the camera apparatus, specifically its capacity to exile the human eye. Smithson notes how we trade a kind of 'cosmic punishment' for the automatic image.[13] The camera is a 'portable tomb' that produces blind spots or creates false mirrors, reflecting things back to us in the manner of

9 Smithson, *Collected Writings*, 146.

10 Smithson's key concepts of site and non-site are based on the relationship between the uncooked quality of the environment or world (site) and the materials drawn from the site into the gallery or museum (non-site), also articulated as the relationship between periphery and centre.

11 Peter Osborne, *Anywhere or Not at All: Philosophy of Contemporary Art* (London: Verso, 2013), 114–16.

12 Smithson, *Collected Writings*, 74.

13 Smithson, *Collected Writings*, 373.

cheap postcards.[14] Smithson sees the photographic or film image as ultimately dislocating and dizzying, 'a camera's eye alludes to many abysses'.[15] Although a medium of light, film according to Smithson would lead the moviegoer into 'a vast mud field of images forever motionless.'[16]

Lost in Film

Smithson's most memorable image and work, *Spiral Jetty* is analysed through the lens of his eponymous film and through the experience he describes of making it, specifically of being lost in reels of film. He describes this feeling of being dislocated in the web of celluloid as a sort of 'lucid vertigo': 'The movie recapitulates the scale of the Spiral Jetty. Disparate elements assume a coherence. ... Adrift amid scraps of film, one is unable to infuse into them any meaning, they seem worn-out, ossified views, degraded and pointless, yet they are powerful enough to hurl one into a lucid vertigo.'[17]

What Smithson means by a 'lucid vertigo' is far from explicit, though there is often a dialectical tension in his work between disembodied abstraction and material reality, the symbolic elements of the non-site and the tangible qualities of the site. Vertigo's mixed condition of falling and lightness also captures this dialectic. Smithson also sees vertigo as a strange portal state induced by film illusion and the endless peddling of space in Hollywood. However, his description ('lucid vertigo') can be specifically relayed to his description of the twister in *The Wizard of Oz* (1939) as he focuses on the image and meaning of the twisting, helical journey in the film.

> I mean it's like to get back to that metaphor of Oz ... through the force of the twister, you're propelled to this central image ... The people go there, the child and the scarecrow, to the Emerald City of Oz, which is a palace but essentially a crystalline buildup

14   Smithson, *Collected Writings*, 121.
15   Smithson, *Collected Writings*, 373.
16   Smithson, *Collected Writings*, 142.
17   Smithson, *Collected Writings*, 152.

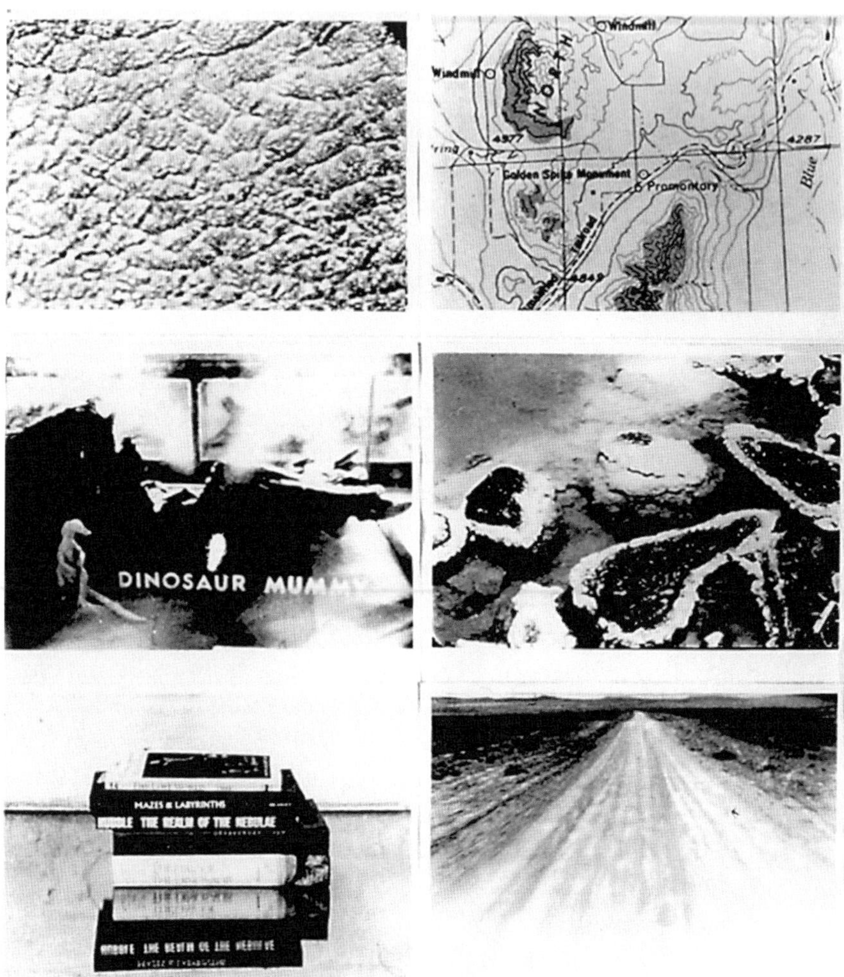

Figure 19a. Robert Smithson, *Stills from the Spiral Jetty (Panel A)*, 1970.

... the terrible aspect behind the element of the twister, behind the centrifugal aspect of the twister, that's a kind of disruption of the gravity almost, or a kind of agony of gravity within the force of the cyclonic centrifugal notion ... I mean the energy gets so intense that it breaks into imaginative, or fairytale results. Like the ultimate reality, its like going from the black and white film in picture to Technicolor.[18]

18   Smithson, *Collected Writings*, 226–7.

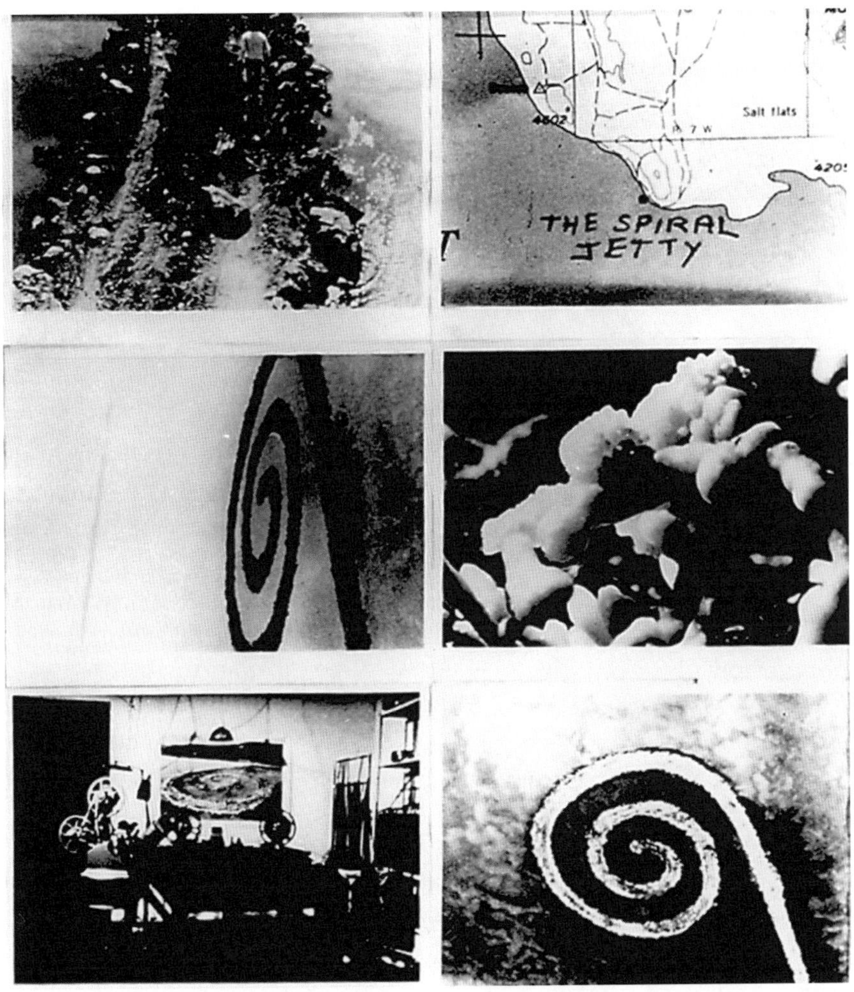

In this interview, Smithson elides gravity and fantasy through the medium and metaphor of film, specifically through *Oz*. His description of the twister connects to key aspects of his work, in particular the strange image of the 'immobile cyclone' ghosted within the site of *Spiral Jetty*, as well as in the spiralling helicopter filming the jetty from above, a symmetry created between the helical motion above and the collapsed spiral below. Seeing the twister's force as gravity's 'agony' also relates to his dialectical vision expressed throughout his art and writing.

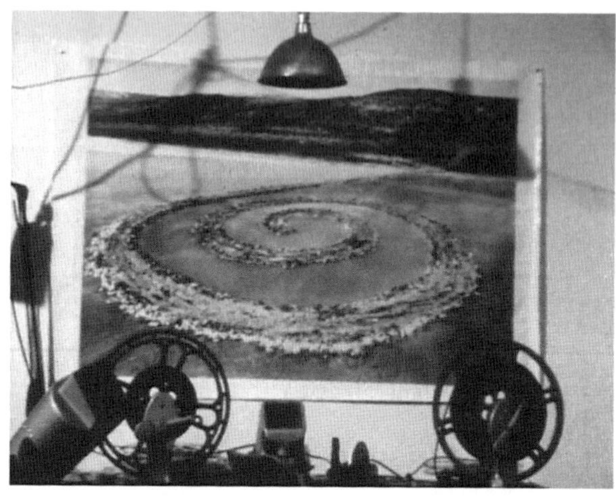

Figure 19b. Robert Smithson, *Stills from the Spiral Jetty (Panel A)*, 1970.

As Smithson articulates things, the power of the twister's rotational movement also grants entry into an illusory world with Dorothy transported from black and white Kansas into Technicolor Oz. Filming *Spiral Jetty* from a helicopter, Smithson accentuates the feeling of vertigo by circling around and moving across the earthwork at different angles and tilts. This feeling of disequilibrium is enhanced by the way light flickers across the frame into what he describes as a 'stomach-turning' experience so that a 'geologic fault' groaned within him.[19] It is also evident from 'A Cinematic Atopia' that the idea of vertigo can be applied to cinematic experience as a whole, namely the disorientations, translocations and temporal jumps fundamental to both making and watching film; furthermore, such dislocations of time and space were critical to Smithson's key dialectic of site and non site, refracted in the relationship between gallery space and the external landscapes of his earthworks. Confusions in space and scale described in the film provide a meditation on the spatial oblivion that cinema creates.

19   Smithson, *Collected Writings*, 148.

Peter Osborne considers Smithson's use and reading of film as part of his overall artistic negotiation of category, medium, and concept. His film of the earthwork complicates the fixed notion of medium since the script is also included in the essay of the same name. Smithson becomes the actor, running around the spiral earthwork or being filmed in helical sweeps from above. As discussed in the previous chapter, cinema is a cross-contaminated art nourished on literature, theatre, and photography. For Osborne, Smithson models the meltdown of category and medium within art 'on a certain cinematic experience of film.'[20] In 'A Cinematic Atopia', he equates watching movies with drowning in 'a vast reservoir of pure perception' or to being faced with 'inventories of limbo' or the 'sprawl of entropy'.[21] Film's condition was spatially dialectical, taking the audience away to distant landscapes, whilst pinning them down in the dark, or moving celluloid strips between camera and projector. He describes the experience of watching film as being 'swallowed up in a morass'.[22] Like his references to swamps and bogs, the ground giving way beneath the audience mediates the sensation of vertigo. Cinematic reels are spinning coils of time, just as geological time shifts the ground like an attack of vertigo slowed into space and mass.

Smithson's essay also exposes a key juncture in his practice between large-scale earthworks and film. It was originally commissioned for Annette Michelson's special issue on film for *Artforum* with a series of stills from his *Spiral Jetty* film included in the essay.[23] 'A Cinematic Atopia' remains his most potent meditation on cinema, revealing the accumulated experience of a young American inevitably nurtured on a diet of Hollywood films and subject to the *cinephilia* of his artistic peers ('The movies give a ritual pattern to the lives of many artists and this induces a kind of "low-budget" mysticism ...').[24] His essay concludes with the startling idea of a film being

20   Osborne, *Anywhere or Not at All*, 100.
21   Smithson, *Collected Writings*, 138–42.
22   Smithson, *Collected Writings*, 138.
23   Robert Smithson, 'A Cinematic Atopia' in *Artforum*, ed. Annette Michelson, 10: 1 (Sep. 1971), 53–5.
24   Smithson, *Collected Writings*, 16.

shot and screened in a dark cave or an abandoned mine, which echoes the whole experience of watching film: 'Time is compressed or stopped inside the movie house, and this in turn provides the viewer with an entropic condition. To spend time in a movie house is to make a "hole" in one's life.'[25] Disused mines and industrial wastelands speak of disorientation and limbo, qualities Smithson recognizes in the nowhere and non-place of cinema. Smithson's selection of a mine or cave rather than a room or gallery replaces the reassuringly rectilinear with collapsed walls or crumbling floors that disrupt the ideal and abstract space of a room. Darkness in the cave mimics the temporal and spatial limbo of the cinematic borderlands and the dark interior of the movie house.

Smithson's reading of film and photography is subtended by allusions to a Gnostic sensibility, what he terms his 'abysmal speculations'.[26] Like Beckett (whose work *The Unnameable* he references at the end of the essay for *Spiral Jetty*), Smithson appropriates the non-transcendent and anti-idealist meanings of Gnosticism. This is particularly evident when he suggests that the sky and sun are not natural vehicles for transcendence, but are instead adamantine ceilings locking us into a monstrous *mise en abyme*. Mirrors locked together, lost vanishing points, suspended gravity and arrested flow become representations of vision sealed up. The action of mass, often described as a creeping crystalline accretion, only goes to prove the existence of a pitiless demiurge, hinted at in 'Incidents of Mirror-Travel in the Yucatan' (1969): 'In the rear-view mirror appeared Tezcatlipoca – demiurge of the "smoking-mirror".'[27] Smithson's mirror displacements concern precisely a degradation of vision as transcendent possibility with light blocked out by the dirt and debris left on the mirrors.

Celluloid, slides, mirrors and angled viewpoints, evident throughout Smithson's writing and practice, confuse vision and bring on attacks of giddiness, disorientation and blind spots. Dislocation is also the keynote of cinema and Smithson somehow moves interchangeably between the renewable horizons of the Hollywood Western and the large earthworks he creates. He translates what he sees as the no-man's land of film through

25   Smithson, *Collected Writings*, 17.
26   Smithson, *Collected Writings*, 373.
27   Smithson, *Collected Writings*, 120.

the film medium itself, through its dissolves and fades. He filters the dislocated spaces of his art practice through the imagery of tired Hollywood scenarios: 'a dim landscape of countless westerns. Some sagebrush here, a little cactus there, trails and hoof-beats going nowhere' or later 'tangled jungles, blind paths, secret passages, lost cities', such fragments of space and horizon as would lead only to 'an abyss within us'.[28] Landscape is a pathetic fallacy for bodily states. In his description of the slate quarry at Bangor-Pen Angyl, Pennsylvania, Smithson notes how, 'countless stratographic horizons ... had fallen into endless directions of steepness. Syncline (downward) and anticline (upward) outcroppings and the asymmetrical cave-ins caused minor swoons and vertigos.'[29] Experiences of vertigo are small demonstrations of entropy read through the body, half falls that reflect the slow collapse of matter under gravity's sway.

As stated, Smithson relates his descriptive term 'lucid vertigo' to the image and mechanics of rotation both in his earthworks and also within the film medium and process. Whirling motion finds complementary expression in his description of film as 'a spiral made up of frames'.[30] He insists on disorientation in the film of *Spiral Jetty* by droning the co-ordinates of the compass, an allusion to Hitchcock's *North by Northwest* (1959), which itself referred to a co-ordinate that does not exist. Smithson includes a rich blend of other cinematic references in *Spiral Jetty*, which have been variously attributed to films such as Chris Marker's *La Jetée* (1962), Hitchcock's *Vertigo* (1958), *Psycho* (1962) and Stanley Kubrick's *2001: A Space Odyssey* (1968).

In an edition of *Artforum* from February 1969, Annette Michelson had already raised the question of profound disorientation in her analysis of Kubrick's *Space Odyssey*.[31] Smithson's earthwork and film were made during a short period (1970–1972), which was marked by the production of key science fiction films, including the Kubrick epic. Smithson references the pulsing red eye of the HAL 900 computer in Kubrick's film by

28   Smithson, *Collected Writings*, 138–41.
29   Smithson, *Collected Writings*, 110.
30   Smithson, *Collected Writings*, 148.
31   Annette Michelson, 'Bodies in Space: Film as Carnal Knowledge', *Artforum*, 7 (6) (February 1969), 54–63.

his use of a single red colour filter to shoot the *Hall of Dinosaurs* episode, described as follows:[32]

> An interior immensity spreads throughout the hall, transforming the light bulbs into dyeing suns. The red filter dissolves the floor, ceiling, and walls into halations of infinite redness. Boundless desolation emerged from the cinematic emulsions, red clouds, burned from the intangible light beyond the windows, visibility deepened into the ruby dispersions.[33]

Colour dissolves rectilinear anchors in space. For Michelson, Kubrick's *2001* had produced a radical shift in viewing conditions at both a formal and thematic level, equated with the philosophical insights of Maurice Merleau-Ponty's phenomenology. She invokes the philosopher's thought that 'All perception is movement. And the world's unity, the perceiver's unity, are the unity of counterbalanced displacements.'[34] In her account of watching Kubrick's *2001*, Michelson writes: 'Once the theatre seat has been transformed into a vessel, opening out onto and through the curve of a helmet to that of the screen as into the curvature of space, one rediscovers, through the shock of recognition, one's own body living in its space.'[35] Kubrick's use of single strip 65-mm Super Panavision and the giant centrifuge set broke new ground in creating cinematic visions of oceanic space and illusions of weightlessness, though the artificial gravity produced by the movement of the centrifuge was to stabilize the body in space, preventing the nausea suffered by astronauts. By Michelson's account, the images of the astronauts floating in space could be seen as emblematic of recent phenomenological understandings of the body's perception, its continual negotiations with uprightness or balance. According to Michelson, by

32   Diana Thater, 'A Man Becomes Unstuck in Time in the Film That Became a Classic!' in *Robert Smithson Spiral Jetty: True Fictions, False Realities*, ed. Lynne Cooke and Karen Kelly (Berkeley and Los Angeles: University of California Press, 2005), 168–70.
33   Smithson, *Collected Writings*, 152.
34   Michelson, 'Bodies in Space', 59.
35   Michelson, 'Bodies in Space', 58.

destroying the given absolutes of floor, wall, and ceiling, Kubrick was able to 'make Keatons of us all.'[36]

Machines that Move

Recent scholarship has taken a spiral turn. In *The Cylinder: Kinematics of the Nineteenth Century*, Helmut Müller-Sievers explores the power of cylindrical motion in nineteenth-century culture and technology.[37] In a more recent publication, *Spirals: The Whirled Image in Twentieth-Century Literature and Art*, Nico Israel proposes the spiral motif as modernist device and trope in works by Duchamp, Joyce, Beckett, and Smithson, acknowledging how the silent movement of reel-to-reel technology haunts his research.[38] Roger Caillois categorizes vertigo as a form of play related to *ilinx*, the Greek word for whirlpool connected to the altered perception experienced on rollercoasters and fairground rides.[39] Film historian, Tom Gunning's general thesis is that early film took on the effects and novelties of fairground and circus in an unsettling of the body's stability. As discussed in the previous chapter, cinema and circus performances are founded on repressions of ground. Both entertainments rely on a shadowing of the ground with dust turned and sand kicked up to sustain the general illusion of things moving with sparse reference to the ground.

From its origins, the film medium was significant in disorientating audiences by severing connection with the ground in fundamental ways. During its early period, cinema was merely a 'deaf and dumb' show

36   Michelson, 'Bodies in Space', 60.

37   Helmut Müller-Sievers, *The Cylinder: Kinematics of the Nineteenth Century* (Berkeley and Los Angeles: University of California Press, 2012).

38   Nico Israel, *Spirals: The Whirled Image in Twentieth-Century Literature and Art* (New York: Columbia University Press, 2015).

39   Roger Caillois, *Man, Play and Games*, tr. Meyer Barash (Urbana and Chicago: University of Illinois Press, 2001), 23.

constructed out of light without even sound to register impact and weight. The way that film cuts the body into pieces is a significant element in the fantasies that it generates, including dreams of flight and fears of falling. Images of bodies moving without reference to their feet or to the ground became a normative condition of film with its singular capacity to move and disorientate audiences. It is precisely cinema's ability to disconnect bodies from the ground with its dolly or crane shots or dissolves that is so often critically dissected in the modern art gallery. Film filters such anxieties about spatial situations with its dolly-slide projections into space or grip-pull shots that suck space into us. Its dissolves and cross fades are phantom experiences of a dizzy turn. Film and its technologies are the machines that move us, tying us to the 'rotor' ride of illusion.

Smithson's reading of cinema is based on the physical mechanics of analogue film as built out of rotation and frames falling through light, as well as the larger movement of frames on a reel in what he describes as 'a spaceless limbo on some spiral reels'.[40] In the late 1960s, Richard Serra had started to expose this physical quality of film frames falling through light in *Hand Catching Lead* (1969), interrogating the process and medium qualities of film. This 'dynamic of falling' is evident in Catherine Yass's final images from the Harold Lloyd series. She adds striations across the image of Lloyd clinging to the clock to indicate the fall of celluloid and its damage as it is pulled through a projector repeatedly. In many ways, Serra's process films aped the silent language of early cinema with grainy black and white images and static camerawork. For Smithson, cinema is an endless burial of image falling upon image, flickering between light and dark in a process of entropic sedimentation. He mirrors this in the pages that fall like dead leaves like 'Earth's history' in the film for *Spiral Jetty* or in the dumping of rocks to make the work.[41]

In terms of Hollywood cinema, the twin motors of falling and rotation are reflected in the patterns exploited in cinematic spectacle: dancers funnelling the air, the 'seductive spirals of vamps' and the kaleidoscopic shapes

40   Smithson, *Collected Writings*, 150.
41   George Baker, 'The Cinema Model' in *Robert Smithson Spiral Jetty*, 83.

of machine-like dancers in the Busby Berkeley fantasies.[42] Hyperkinetic, gravity-defying bodies of mainstream Hollywood films and comics such as Buster Keaton, Fred Astaire or Superman reflect a kind of cultural 'will to power'. Being hubristic spectacles of discharged energy, they possess the power to bewitch or bore us. The Fred Astaire RKO musicals of the 1930s were often set in *luxe moderne* interiors with gleaming white surfaces, mirrors and polished Bakelite floors. Robert Smithson wrote about the quality of such *ultramoderne* buildings with their mirrored interiors and bewildering symmetries, setting out a dark vision to unsettle their surface illusionism. Again, he makes explicit reference to the ideas of Gnosticism: 'The 'thirties recover that much hated Gnostic idea that the universe is a mirror reflection of the celestial order – a monstrous system of mirrored mazes. The 'thirties became a decade fabricated out of crystal and prisms, a world heavy with illusion. Never has the phantasmal appeared so solid.'[43]

Smithson equates mirrors with film: 'as everybody knows the mirror is a symbol of illusion, as immaterial as a projected film.' RKO's designer, John Harkrider based the Silver Sandals rooftop nightclub in the Astaire musical, *Swing Time* (George Stevens, 1936), on the 1934 Rainbow Room, which was installed with a revolving dance floor and a glass ceiling that mimicked a starry night. Gleaming *moderne* interiors were seen as apt contexts for Astaire's gravity-defying dances. Smithson's reading produces a shocking reversal by recognizing a different type of illusionism at work: 'The Ultramoderne puts one in contact with vast distances, with the ever-receding square spirals; it projects one into mirrored surfaces or into ascending and descending states of ludicity. Walls, rooms and windows take on a vertiginous immobility – Time engulfs space.'[44] Smithson unveils the seductions of the eye occasioned by cinema or the glittering interiors of the Deco building. If we inhabit a Gnostic world, then the interlocking and symmetrical mirror maze of the *ultramoderne* building is precisely that artificial

42   Hillel Schwartz, 'Torque: The New Kinaesthetic of the Twentieth Century' in *Incorporations*, ed. Jonathan Crary and Sanford Kwinter (New York: Zone Books, 1992), 101.

43   Smithson, *Collected Writings*, 64.

44   Smithson, *Collected Writings*, 65.

hell. In addition to his mirror displacements, Smithson's *Enantiomorphic Chambers* (1965) encapsulate a similar idea with two wall-mounted mirrored chambers that are placed in an ambidextrous relationship to each other and produce retinal confusion by disrupting stereoscopic vision.[45]

Hollywood culture had always been fired up by psychotic energy and city rhythms, set in motion by the camera. Vertigo and turbulent motion is signed as both a metaphor and a real hallucination of motion, betraying an underlying bad faith over society's rapture with energy. Smithson retards these rotations and uncoils the energy of such powerful helixes in an undoing of film illusion. Just as time overwhelms the glittering interior of the *moderne* interior, so film is inevitably a false mirror of eternity.

Film's mechanics of rotation is mirrored in the cinematic experience: life and time itself turn before our very eyes or we are turned in cinematic space. It is therefore unsurprising that film directors often used turbulent motion or spinning as a metaphor for film itself. Soviet Director, Dziga Vertov coined his name from the Russian term for 'spinning top'. One thinks also of François Truffaut's 'rotor' ride in *The 400 Blows* (1959). When the young 'Antoine' steps into the fairground machine, he blends into the machine as it spins. Anne Gillain explains how in Truffaut's film, 'the rotor resembles early machines that allowed an illusion of movement to be created from a rapid succession of fixed images.'[46] Mechanical rotation is also evident in Alfred Hitchcock's spinning or spiral motif in *Vertigo* (1958) and in Stanley Donen's tumbling hotel room used for Fred Astaire's 360-degree dance in *Royal Wedding* (1951). These are literal transcriptions of rotation bound up with the cinematic process itself.

Smithson recognizes this propensity to spinning in Hollywood cultural forms and hyperbolic bodies but his interest in spiralling forms was to retard them into a state of collapse. Certain film images seem to have greater potency than others. In his essay, 'A Cinematic Atopia', the artist reveals his leaning towards the 'lurid sensationalism' of Hitchcock and singles out 'the

45 Ann Reynolds, 'Enantiomorphic Models' in *Robert Smithson* (Los Angeles: Museum of Contemporary Art and University of California, 2004), 137–40.
46 Anne Gillain, *François Truffaut: The Lost Secret*, tr. Alastair Fox (Bloomington: Indiana University Press, 2013), 30–1.

shot in *Psycho* where Janet Leigh's eye emerges from the bathtub drain after she's been stabbed.'[47] Eyes and blinded vision also seem to hold particular meaning for the artist. In his *Spiral Jetty* essay, Smithson alludes to the sun shining red through his eyelids with the entire landscape engulfed in crimson light like the blood circulating in his eyes and writes: 'My movie would end in sunstroke.'[48] Again, the sensation of dizziness is fundamental to a loss of spatial parameters and is an assault on the steadiness that vision generally supplies. 'Lucid vertigo' in Smithson's thought and practice is evocative not only of the effect of the film illusion, but of a kind of sunstroke imagined by Georges Bataille in his ghastly image of the pineal eye:

> At the same time, this ocular tree is only a giant (ignoble) pink penis, drunk with the sun and suggesting or soliciting a nauseous malaise, the sickening despair of vertigo. In this transfiguration of nature, during which vision itself, attracted by nausea, is torn out and torn apart by the sunbursts into which it stares, the erection ceases to be a painful upheaval on the surface of the earth and, in a vomiting of flavorless blood, it transforms itself into a vertiginous fall in celestial space, accompanied by a horrible cry.[49]

Light mutilates vision and stability, propelling us into a dizzying spin and eroding our sense of ground as organized into selfhood. The image of the mutilated eye has pervaded cultural thought since *Un Chien Andalou* (1929) and is echoed in Smithson's allusions to Janet Leigh's dead eye, HAL's inorganic computer eye glowing red, and the swirling spirals across Kim Novak's eye in Saul Bass's title sequence for *Vertigo* (dir. Hitchcock, 1958). Recounting his experience of the site for Spiral Jetty, he describes 'a floating eye adrift in an antediluvian ocean.'[50]

47   Smithson, *Collected Writings*, 139.
48   Smithson, *Collected Writings*, 148.
49   Georges Bataille, *Visions of Excess: Selected Writings 1927–1939*, ed. Allan Stoekl, tr. Stoekl with Carl L. Lovitt and Donald M. Leslie Jr (Minneapolis: University of Minnesota Press, 2004), 84.
50   Smithson, *Collected Writings*, 148.

Other Turns

Contemporary artists often disrupt the illusions created by film by re-connecting the body with the ground or viewing space within a gallery, making the apparatuses of illusion, the cameras and projectors, continuous with the audience's space. By deconstructing film's mechanics and projection, they emphasize the conditions of the projection space. Projectors and 16-mm films are a common sight in the contemporary art gallery with the mechanics of film moved centre stage or purposively exposed, noisily whirring as the celluloid drops through the machine. This visibility was given baroque extension in Simon Starling's *Wilhelm Noack oHG* (2006), shown at the Venice Biennale (2009), in which the celluloid of the accompanying film about a Berlin metal factory moves in extravagant loops through a large spiral sculpture, itself constructed in the same metal works. By unveiling the structural parameters and projection spaces of film, artists unlock the real agencies of mechanical production and reproduction within cinema. Furthermore, artists select, appropriate, and manipulate mainstream Hollywood film to explore cinema's commodification of vision, as explored in exhibitions such as *Spellbound: Art and Film* at the Hayward gallery in 1996 (curated by P. Dodd and I. Christie, 22 February to 6 May 1996) or *Notorious: Alfred Hitchcock & Contemporary Art*, shown at Modern Art Oxford in 1999. I focus here on how contemporary artists have examined the spinning mechanics of film to uncover how the reels, gyres, and helixes that drive industrialized sound and vision are uncoupled from their hidden mechanics.

In considering the specific rotations that support film and the mechanics of industrialized entertainment, I should take account of other contemporary artists, who have excavated the relationship between technology and illusion. More recently, artists such as Martin Kersels, Steve McQueen, and Rodney Graham have unravelled qualities of the kinematic. Californian artist Martin Kersels's work *Pink Constellation* (2001) was inspired by Fred Astaire's 360-degree dance in Stanley Donen's *Royal Wedding* (1954). Intrigued by the illusion of Astaire dancing effortlessly on walls, ceiling and floor, he re-created the technology, blending references to *The Wizard of Oz*, Buster Keaton, and Astaire in a catastrophic slow-motion dance. Kersels is a large man in height and girth, measuring 6 foot 7 inches and weighing in at around 360 lbs. In *Pink Constellation*, his larger-than-life dimensions are made momentarily weightless as the tumble-room rotates. In a witty parody of Astaire's easy grace, Kersels is ill matched to the environment of a teenage girl's bedroom, replete with pink walls and pop posters. Within the one-camera video, dancer Melinda Ring enters the tumbling bedroom, pirouetting and twirling gracefully within the centrifuge, her fragility contrasted with the lumbering frame of Kersels. James Trainor cites the popular cultural references of the installation at Deitch Projects (New York) in 2001:

> As a Southern Californian, Kersels draws from a rich collection of associations: suburban teen culture, earthquakes, natural disasters and the omnipresent movie industry. Indeed, an accompanying video of the artist and a young girl taking turns inside the gyrating room, *Pink Constellation* (2001), pays low-budget homage to the tumbling house in movies, from Buster Keaton's 1921 classic, *The Boat*, to Judy Garland's tornado-twirled farmhouse in *The Wizard of Oz* (1939) and Fred Astaire's 360° tap-dancing promenade around a hotel room in *Royal Wedding* (1951).[51]

Tumbling rooms and houses evoke the silent films of Buster Keaton, as well as the spaces of early Hollywood with its rickety sets, flimsy stages, and wooden houses with no roof or fourth wall. Peter Conrad describes these sprawling locations of early film: 'A society has unpacked itself hurriedly from the moving van, which perhaps – stranded between the desert and the ocean – came to the wrong address. The roads have not been sealed; the trees have not grown …'[52] Creating the illusion of weightlessness in tumble-room technology involves fixing a stable camera and fixing down the furniture and props of the room, whilst the bodies are weightlessly suspended in the tumble-room rotation. This technology, used by Stanley

---

51   James Trainor, 'Martin Kersels' in *Frieze*, 60 (2001) <https://frieze.com/article/martin-kersels>, accessed 4 April 2016.

52   Peter Conrad, *Modern Times, Modern Places: Life and Art in the 20th Century* (London: Thames and Hudson, 1999), 424.

Donen in *Royal Wedding* (1951), was essentially the same as that used later by Stanley Kubrick for the rotating space station in *2001* on a much larger scale.

Previously, in 1998, Steve McQueen combined a pink-walled installation and the mechanics of rotation in a sculptural installation called *White Elephant*. *White Elephant* is a circular sculpture akin to a children's roundabout, designed to reflect the exhibition audience in its highly polished chrome surface and everything else in the room, including the shocking pink walls. First installed as part of an exhibition of McQueen's films from the 1990s, the title, *White Elephant*, referred partly to this work's own peculiarity within a film exhibition and to the obsolete technology of the zoetrope.[53] The etymology of *zoetrope* is from the Greek meaning of 'life turning'. In this work, McQueen invites the audience to spin the sculpture in order to dissolve the space into a dizzying blur, creating a sensation akin to vertigo experienced on the playground roundabout.[54] McQueen examines the interface between illusion and movement, intrinsic to the process and reception of film by the way his roundabout 'structures the terms of the production and reception of the illusion'.[55] By spinning the roundabout, the audience creates the action and illusion of a large-scale zoetrope.

In contrast to mechanical rotation, natural spinning motion is evoked in other examples of popular film. In addition to the *Wizard of Oz* twister described by Smithson, another key reference is the tornado that appears in Buster Keaton's *Steamboat Bill Jnr.* (1928). In Keaton's film, the twister rips through the town, collapsing buildings and whipping up other debris along with his body. Keaton is made helpless and often weightless in the face of the obliterating storm as the flimsy Hollywood set is shredded into confetti. In *Deadpan* (1997), McQueen reimagines the scene of the falling house in Buster Keaton's *Steamboat Bill Jnr*, taking on Keaton's role with the timber wall that falls repeatedly around him in an extended, purgatorial

53 Okwui Enwezor, 'From Screen to Space: Projection and Reanimation in the Early Work of Steve McQueen' in *Steve McQueen: Works 1993–2012* (Heidelberg: Kehrer Verlag, 2013), 32.
54 Enwezor, *Steve McQueen: Works*, 33.
55 Enwezor, *Steve McQueen: Works*, 33.

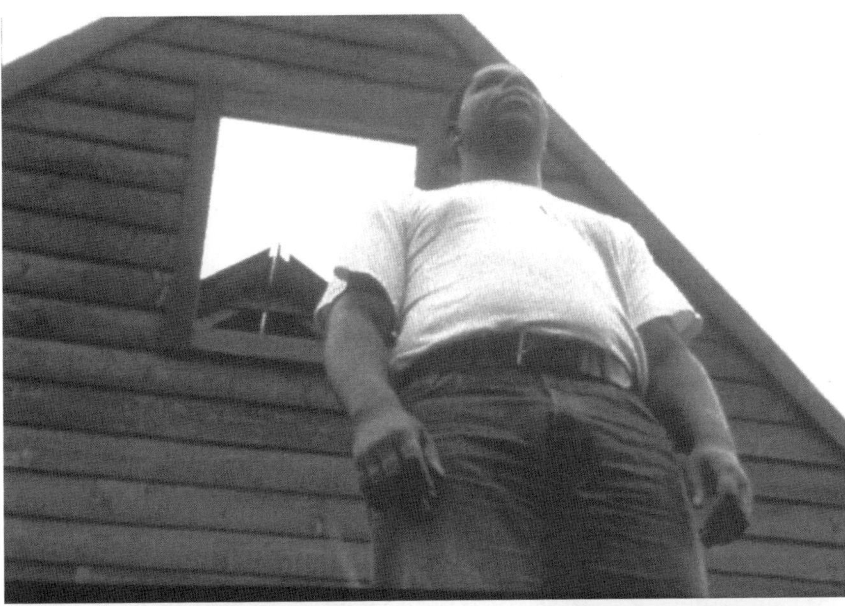
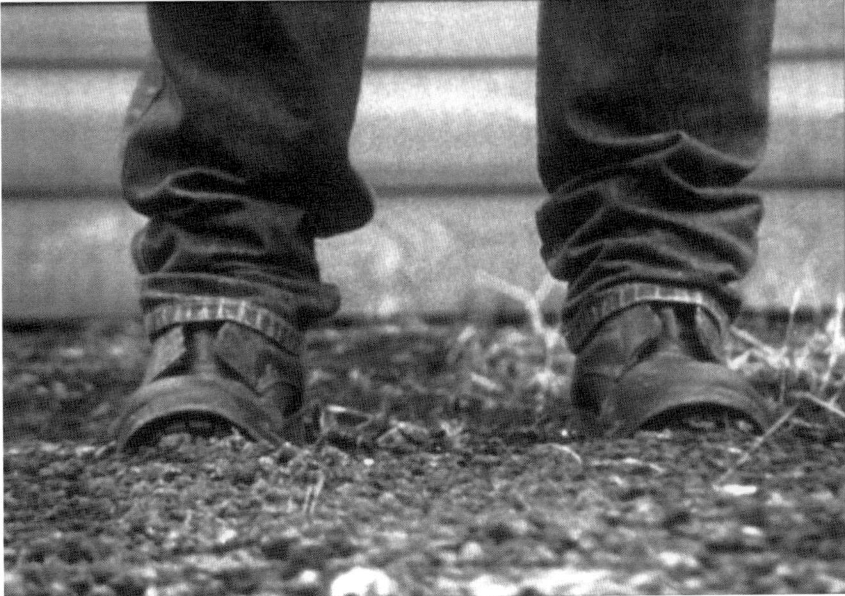

Figures 20a–20b. Steve McQueen, *Deadpan*, 1997.

gag.[56] He montages shots of his deadpan expression, of his boots (notably lace-less to hint at some dark history of the slave or sharecropper) and the wooden wall falling around him on a continual 4-minute, 35-second loop that deepens and prolongs the torment. He creates a counterpoint between horizontal and vertical planes, using his own upright body, which is apparently unperturbed by either the threat or event of the falling wall. As it falls, the wall's trajectory downwards creates flickering between light and dark as if shadowing the action of the film itself. As in Richard Serra's *Hand Catching Lead* (1969), the film action mirrors the action of film, as frames falling through light.

In another work called *Drumroll* (1998), McQueen placed three cameras inside a large oil drum, rolling the drum through the streets of New York, wearing a vivid pink jacket akin to a circus ringmaster's. The resultant 25-minute film is largely made up of a blur of images captured by the camera turning along the pavement with McQueen's voice audible, as he asks pedestrians to move out of his way. One key effect of cinema is to sever the connection between body and ground, producing what Rudolf Arnheim describes as that 'curious gliding, floating character'.[57] By reconnecting the camera with the ground, McQueen disrupts the cinematic magic of levitating or flying. Okwui Enwezor summarizes: 'Drumroll is an essay on the action of the camera, which it reduces to a cinematic form of automatism, an idea with precedents in the Structuralist films of the 1960s. The viewing of the film elicits a certain ungrounding, for the spinning camera, caterwauling down the street, produces a dizzying visual nausea.'[58] McQueen fuses motion and camera mechanics in such a way as to unlock the very foundational language of film. He uncovers its very structure in terms of production and reception in what Enwezor describes as 'still born'

---

56   *Deadpan* was rescreened at 44, Times Square, New York, in 2009, with McQueen's impassive face staring out amidst a sea of garish advertising. Rescreening *Deadpan* also prompted the artist to reconsider the film's political implications post-Katrina.

57   Rudolf Arnheim, *Film as Art* (Berkeley and Los Angeles: University of California Press, 1957), 116–17.

58   Enwezor, *Steve McQueen*, 30.

cinema. Early cinematic toys such as the zoetrope and phenakistiscope required active participants, willing to operate and control their mechanisms to conjure the illusion. Marcel Duchamp's *Rotary Demispheres* (1924) is based on the early phenakistiscope (1833) and his interest in spinning and optical illusion is evident in his other incursions into optical play, such as the *Rotoreliefs* and his short film made with Man Ray and Marc Allégret, *Anémic Cinéma* (1926). By recovering these proto-cinematic forms, artists also recover the physical and material constituents that support film's play of light and illusion.

In Rodney Graham's *Phonokinetoscope* (2001), the audience also become part of the projection process by placing or re-placing the needle onto a turntable in order to start the film up. Graham creates a pun on the idea of a 'trip', both hallucinogenic and physical. Taking a lead from Thomas Edison's invention of the phonokinetoscope, Graham's film installation links the action of a turntable and vinyl record with the 16-mm film loops of a bicycle ride undertaken by the artist in Berlin's famous *Tiergarten*. His trip on a bike alludes to, amongst other things, Bas Jan Ader's famous *Fall II*, the short film in which the Ader rode his bike nonchalantly into an Amsterdam canal, as discussed in Chapter 2.[59] Ader's renunciation of control to the agency of gravity is evident in Graham's understanding of the swerve as a moment of uncontrollable deviation, as well as in his use of the consciousness-altering substance, LSD. Again, a paradigm of looping emerges through a series of moves in which the spinning turntable, spinning bicycle wheels, and coiling spools of 16-mm film are assembled in Graham's gambit of weaving background, foreground, and audience together. His trip (both physical and hallucinogenic) circling around the Berlin *Tiergarten* is linked by the artist to the viewer's own 'movement on-the-spot', that fundamental experience shaped by cinema.[60]

59   Rodney Graham, 'A Thousand Words' in Shepherd Steiner, *Rodney Graham: Phonokinetoscope* (London: Afterall, Central St Martin's College of Art and Design, 2013), 105.

60   Steiner, *Rodney Graham: Phonokinetoscope*, 42.

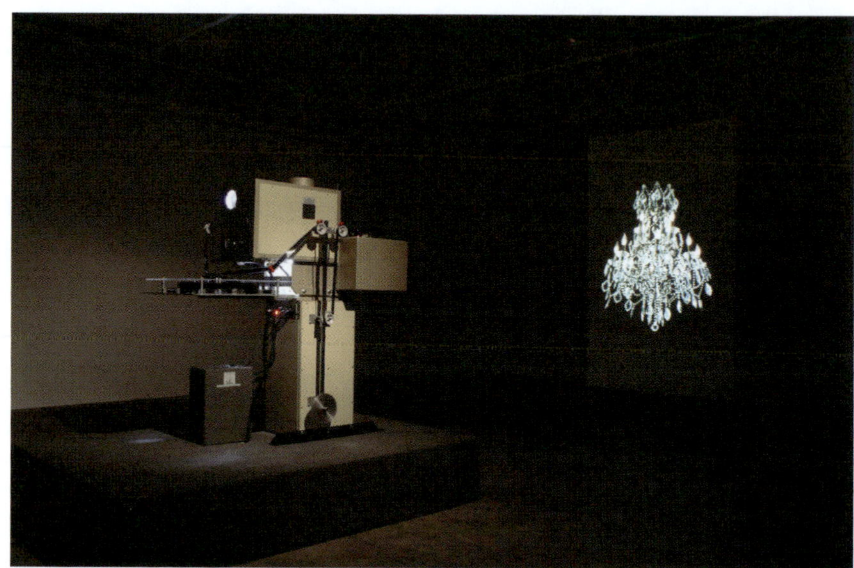

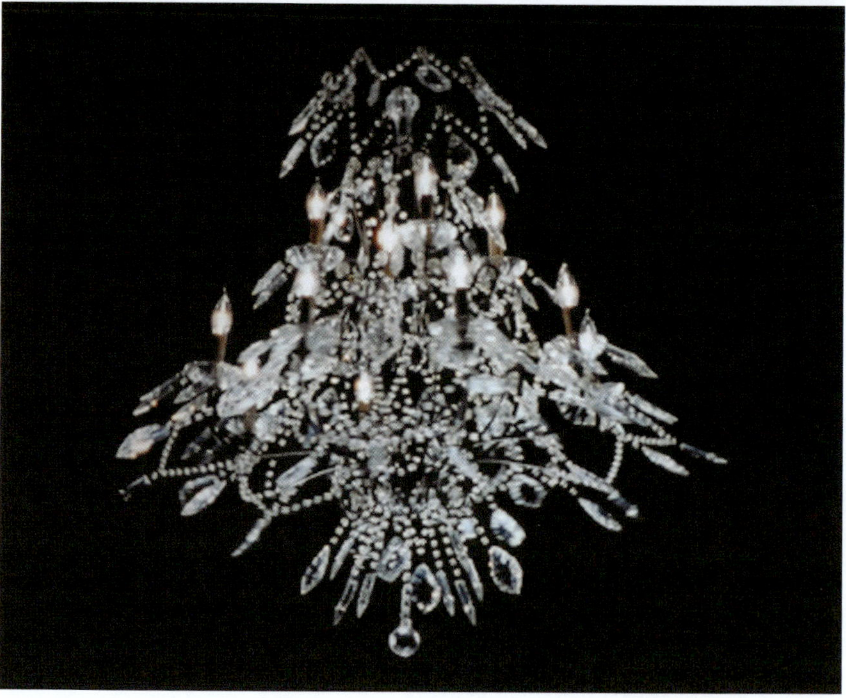

Figures 21a–21b. Rodney Graham, *Torqued Chandelier Release*, 2005.

Rodney Graham's work has often incorporated the structure and associated mechanics of looping and repetition in film. How does the motion of looping and rotating connect with gravity? In a more recent work, *Torqued Chandelier Release* (2005), Graham plays with the laws of dynamics, alluding to Isaac Newton's 1689 spinning bucket experiment. Turning in the darkness, a crystal chandelier rotates slowly then faster. Coming to a halt, it then rotates in the opposite direction. The process of tension-release and effort-exhaustion is calculated to fit on a single roll of film. Again, there is a sense in which he examines the structure and materials of film within the model of an early physics experiment. In *Vexation Island*, he had faintly echoed Newton's apple into his falling coconut sequence. Here Graham makes his inspiration for *Torqued Chandelier Release* more explicit:

> Issues pertaining to Newtonian gravity and rest vis a vis large and rapidly moving lighting fixtures were graphically brought to my attention at a young age when I witnessed, while viewing the 1952 film *Scaramouche*, the near impalement of Stewart Granger by means of a falling chandelier ...[61]

Graham's explorations into cinema are complexly iterated: film's effects are viewed as an invention that creates both visceral impact and sense-dulling hypnosis. Shepherd Steiner describes Graham's works as a set of 'caricatured resistances to cinema that make up the history of his film works'.[62] In a statement almost reminiscent of Smithson, Graham talks of the discovery of a 'personal substratum of screen memories' that were both 'beautiful and tawdry'.[63] He cites memories of family camping expeditions in British Columbia with his father acting as cook, camp master, and projectionist ('I attended these screenings and thus ... remained in close proximity not only to patriarchal authority but also to the cinematic apparatus itself').[64]

61   Rodney Graham 'Artist Statement, 2005' in *Rodney Graham: Through the Forest*, ed.
      Friedrich Meschede (Ostfildern: Hatje Cantz Verlag, 2010), 136.
62   Steiner, *Rodney Graham*, 69.
63   Graham in Meschede, *Rodney Graham*, 86.
64   Graham in Meschede, *Rodney Graham*, 86.

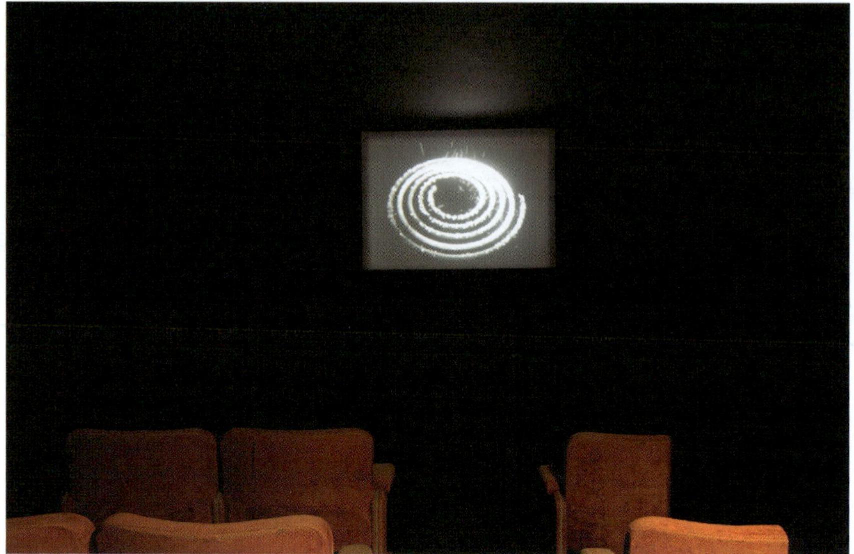

Figure 22. Rodney Graham, *Coruscating Cinnamon Granules*, 1996.

Graham's interest in spiral movement and dizziness recalls Smithson's *Spiral Jetty*. His *Coruscating Cinnamon Granules* (1996) is a short 16-mm film loop projected within a newly installed, kitchen-sized space with eight cinema seats. The title operates as a pun on cinema in 'cinema-cinnamon'.[65] Graham's initial idea for this work was sparked by a family friend, who had spoken of a house-selling tip whereby cinnamon granules would be shaken over a spiral electric hob, and heated in order to cover up any unpleasant house smells. Shot in black and white, the 4-minute loop films the entire process of heating the spiral element to the final extinction of the burning granules. He explains it as 'lighting event' performed at home:

65   Alex Alberro in Steven Harris, 'Pataphysical Graham: A Consideration of the Pataphysical Dimension of the Artistic Practice of Rodney Graham', *Tate Papers 6* (2006) <http://www.tate.org.uk/research/publications/tate-papers/06/pataphysical-graham-consideration-of-pataphysical-dimension-of-artistic-practice-of-rodney-graham>, accessed 4 April 2016.

> [I]ntended to be projected in a cinema bearing the dimensions of my own modest kitchen: a glittering mini-spectacle that resembles a constellation of stars that appear before one's eyes after a mild blow to the head, created by simply spreading granulated particles of the common household spice over the surface of a spiral electric cooking element before turning the element on in darkness.[66]

The artist's use of scale and cosmic reference is a reminder of Smithson's erosion of space, scale and time in *Spiral Jetty*. Graham's dizzy turns are various: a druggy bike ride around the *Tiergarten*, a continually rotating chandelier, the spiral constellation of an electric hob described as a 'mild blow to the head', and his falling-coconut-induced unconsciousness in *Vexation Island* (1997). Such altered states are not uncommon in his films; sleep in *Halcion Sleep* (1994), the idea of 'seeing double' in *Fantasia for Four Hands* (2002), and Freud's uncanny returns configured in so many of his works. He loosely connects the comic idea of falling (suggested by a blow to the head and explored subsequently in *Vexation Island* and *City Self/Country Self*) with the image of seeing a constellation of stars. Smithson's work is metaphorically present in Graham's short-circuit of film, spiral motion, and reference to cosmic scale. Certainly, Graham has discussed Smithson's figure of the spiral (though attenuating its curves to that of a 'meander') in his essay, 'Smithson's Brain'.[67]

> The figure, which to my thinking most adequately circumscribes Smithson's mind and writing, is not the centredness circle, the spiral or the labyrinth, but rather the meander – a figure that appears often enough in his work. It suggests a circuitous journey, and a leisurely one, at least, one guided ultimately by the imperative of the site-seer's pleasure. And the adjective *meandrine* means full of windings, especially of a rather beautiful genus of corals with a surface bearing an amusing resemblance to a fossilized human brain.[68]

66   Rodney Graham, 'Siting Vexation Island', *Island Thought: an archipelagic journal published at irregular intervals* (Brussels: Yves Gevaert Verlag, 1997), 12–13.

67   Rodney Graham, 'Smithson's Brain' in *Robert Smithson in Vancouver: Fragment of a Greater Fragmentation* (Vancouver: Vancouver Art Gallery, 2004).

68   Graham, 'Smithson's Brain', 2004.

This image of brain coral is the visual conceit that combines the large-scale qualities of nature with mental meandering in the notion of consciousness itself. Graham's associations of differing states of semi or unconsciousness provide powerful incursions into questions of agency and will. As explored in Chapter 2, Graham creates his own definition of Lucretius' swerved fall (*clinamen*), describing the loss of subjective agency or will within the natural order: 'The word is from Lucretius, for whom it signifies the sudden and unpredictable swerve of a single atom from its otherwise pre-ordained trajectory ... It is the clinamen, according to the physicist, that breaks the endless chain of fate and yields the law of nature.'[69] The *clinamen* is a doubled-edged sword; it is both the chance swerve that allows for creative reinvention and freedom, but is also defined as the swerve towards an object of obsessive compulsion, which in turn acts against free will. Swerves may signal moments of opportunity to break out of a loop or a habit, whether mechanistic or psycho-behavioural. Bas Jan Ader's bicycle ride into the canal is an inexplicable moment of deviation (or planned compulsive act of falling) that Graham cites in his essay on *Phonokinetoscope* for *Artforum*. 'Bas Jan Ader rode his bicycle into an Amsterdam canal not long after Butch Cassidy came out, and his work is as important to me as Bacharach's soft pop.'[70] In these other turns, Graham, McQueen and Kersels examine the mechanics of industrialized entertainment, unwinding these coils of illusion.

Helical Journeys

The model of movement within a vortex or helix has become an important image of power, transformation and collapse. In *Rings of Saturn*, W. G. Sebald's then home, the flatland of East Anglia, is the context for

69 Rodney Graham, 'Freud's Clinamen' in *Rodney Graham: Vancouver Art Gallery* (Vancouver: Vancouver Art Gallery, 1988), 60.
70 Graham in Steiner, *Rodney Graham*, 105.

an apocalyptic vision of a sandstorm. When the storm of 'sinister spirals' abates into an eerie calm, Sebald's narrator climbs out of his refuge and notes: 'This, I thought, will be what is left after the earth has ground itself down.'[71] For Sebald, history is not linear and historical trauma is a spiral emanation. Similarly, Smithson's use of the spiral is one in which he emphasizes a trajectory of recoil, a spiralling movement out of anthropocentric history. In *Spiral Jetty*, he telescopes the murky ends of prehistory and science fiction. The whirling movement of his imagined tornado connects to his central image of the jetty first intuited as 'an immobile cyclone'.[72] Smithson articulates and seeks to undo the intense energy of the twister (its 'terrible aspect') as a conduit into a world of illusion. *The Wizard of Oz*, the *ur* film about leaving or escape might be seen as a summary of the whole enterprise of Hollywood cinema and the way it peddles illusion.

Artists other than Smithson have explored the natural phenomenon of the tornado. Francis Älys's on-going project, *Tornado*, has been running since 2000. Every March, he drives out southeast of Mexico City and films himself running into the eye of the twister with only his nose and mouth covered. He describes his strange motivation as an attempt to locate a sense of survival and intense immediacy. By forcing his body into the eye of the sandstorm, he films short sublime moments of radical disorientation in a poetics of will and ego loss.

In *Anywhere or Not at All*, Peter Osborne describes certain 'processes of negotiation' in Robert Smithson's seminal work, *Spiral Jetty* (1971) in relation to his film of the work, in which he appears as a 'wild, "hunted" figure, in an exuberant parody of the scene in Alfred Hitchcock's 1959 *North by Northwest* in which the character played by Cary Grant is dive-bombed by a crop-dusting airplane.'[73] Smithson mirrors the stumbling journey of Roger Thornhill, running across the earthwork and intoning its compass co-ordinates and materials. This scene was originally conceived as a tornado that spirals towards Cary Grant, a brute and inexplicable force of destruc-

---

71   W. G. Sebald, *Rings of Saturn*, tr. Michael Hulse (London: Vintage Classics, 2011), 229.

72   Smithson, *Collected Writings*, 146.

73   Osborne, *Anywhere or Not at All*, 105.

tion like the attacking birds in Hitchcock's eponymous horror film. In *Spiral Jetty*, Smithson is filmed from the swirling helicopter, which counters his spiral movements around the jetty in the manner of a hunt. At this point in time, helicopters were indelibly associated with the Vietnam War and their hovering suspension became synonymous with the lack of momentum in resolving the Vietnam War. According to Nico Israel, Smithson's use of the helicopter was adapted to the figure of the helix and to the 'notion of spatio-temporal arrest'.[74] The helicopter's slanting view of the jetty produces a tension between the concealed and revealed, between blind spots and blinding light, underscored by the heavy drone of the machine and the grim incandescence of the scene. In this tumult of vertigo, Smithson shifts and turns the ground beneath as the helicopter shifts across the earthwork below. As in his sinister reading of the mirrored ceiling of the Art Deco interior, which glares back at us, his aerial projects are anti transcendent in meaning.

Contemporary artists such as Melanie Smith, Rodney Graham, Cyprien Gaillard, and Catherine Yass have exploited the slanting flight of the helicopter to explore the ontological and philosophical fault lines in the fantasy of the stable viewpoint. Rodney Graham's *Edge of a Wood* (1999) was a film shot from a helicopter in which the rotor blades whip up the vegetation, illuminating the wood in an unsettling act of surveillance. Contemporary artist Melanie Smith has more recently responded to Smithson's *Spiral Jetty* with her film, *Spiral City* (2002). Filming from a helicopter in broader and higher circles of elevation, Smith emulates Smithson's own aerial film and documents the ever-expanding sprawl of her adopted Mexico City. Seen from above, the city's grid is eroded by structures and buildings heaped on other collapsed buildings in an image of crystalline build up and accumulated mass. Smith makes a parallel between the literal masses of the city's building and the social deprivation of the mega city's inhabitants. Cyprien Gaillard's *Desniansky Raion* (2007), discussed in my first chapter, was filmed from a helicopter. Again, this circular formation of a Kiev housing suburb is likened to an Aztec city or Stonehenge, dropping through layers or 'spiral emanations' of history. British artist Catherine Yass also uses a helicopter flight to capture the instability and displacement of

74   Israel, *Spirals*, 184.

an unmanned lighthouse, 5 miles from the coast of Bexhill. The resultant film exploits spiral movement downwards into the water and around the lighthouse, balanced precariously on a post out at sea. In post-production, Yass inverted some of the images so that the lighthouse is seen upside down. In all these helical flights, there is a dizzying assault on the coherence of a stabile viewpoint in the artists' varying attempts to unlock the social and psychological meanings of architecture, site or history. These artists pierce the layers of illusion and unravel the assumptions of the bird's eye view as providing leverage or a vantage point from which to understand history. In *Spiral Jetty*, Smithson collapses cave man into space man, but also the blinding light from above into the sunken morass below.

Vertigo

What other psychological meanings of vertigo exist? In *Emotional Vertigo*, Danielle Quinodoz outlines the variants of vertigo through the processes of ego differentiation. What she terms, 'fusion-related vertigo' is the most primitive form of vertigo, by which the sufferer feels as if they are about to disappear altogether or fade away. It is felt less as a sensation of falling and more as a complete dissolution of the self into the world. Another form of vertigo is 'expansion-related', an experience of space as devoid of objects, felt as an inability to experience boundaries or limits.[75] This loss of differentiation between ego and environment is fundamental to Freud's reading of the death drive. An attachment to dead things is also the cornerstone of his essay about the uncanny.

Marget Iverson's study of *Spiral Jetty* in *Beyond Pleasure* produces a reading of the earthwork as Smithson's notion of the death drive made visible, argued through his interest in the writings of Anton Ehrenzweig and Bataille. Iverson rehearses the Freudian concept of the death drive

75   Danielle Quinodoz, *Emotional Vertigo: Between Anxiety and Pleasure*, tr. Arnold Pomerans (London: Routledge, 1997), 21.

(*Thanatos*) as a cosmic and psychological model of entropy and tension reduction. She quotes Freud, who claims: 'Protection against stimuli is an almost more important function for the living organism than reception of stimuli.'[76] Freud associates repetition compulsion as a need to bind together and consolidate the ego, in contrast to the unbinding principle of the death drive in its dissolution of the ego's boundaries. Within this model of ego binding and unbinding, Smithson's own mediation of Anton Ehrenzweig's book, *The Hidden Order of Art* (1967) is significant in the claims he makes for the inception of *Spiral Jetty*. His retarded cyclone suggests a state of dedifferentiation after the convulsion of a storm. Iverson draws on Ehrenzweig's understanding of the critical role of dedifferentiation within any creative act: 'In modern art the ego rhythm is somewhat one-sided. The surface gestalt lies in ruins, splintered and unfocusable, the undifferentiated matrix of all art lies exposed ... The pictorial space advances and engulfs him in a multidimensional unity where inside and outside merge.'[77]

The death drive is marked by association with creative acts of dispersal, excretion, and an impulse towards blurred boundaries. Jackson Pollock lurks large in Ehrenzweig's account, because he is able to withstand and contain this pulverizing of form into states of in-distinction. In 'A Sedimentation of Mind', Smithson connects the rhythmic looping of Pollock's painting process to a comment by Jean Dubuffet: 'The revolution of a being on its axis, reminiscent of a dervish, suggests fatiguing, wasted effort; it is not a pleasant idea to consider and seems instead the provisional solution, until a better one comes along, of despair.'[78] Dedifferentiation is characterized by a whirled state of matter, which for Smithson was synonymous with his key conceptualization of entropy and was linked by Ehrenzweig with the death drive. Smithson talks about the quality of entropic dispersal: 'In other words, you're into this area of *de*differentiation that Ehrenzweig talks about where the gestalt becomes something else. The entropic aspect comes in.'[79] Subjective experiences of collapse are akin to feelings of being

---

76 Margaret Iverson, *Beyond Pleasure: Freud, Lacan, Barthes* (Pennsylvania: Pennsylvania State University Press, 2007), 74.

77 Anton Ehrenzweig in Iverson, *Beyond Pleasure*, 77.

78 Smithson, *Collected Writings*, 110.

79 Smithson, *Collected Writings*, 199.

eaten by space. Vertigo arises as the confusion felt by an individual in relation to the surrounding space.

In his essay, 'Mimicry and Legendary Psychasthenia' (originally published in a 1935 edition of *Minatoure*), Roger Caillois's analysis of animal mimicry is related to an experience of 'depersonalisation by assimilation to space.'[80] Caillois observes how an animal's ability to camouflage itself was often a spontaneous or evolutionary blending into the environment. He notes how the stick insect camouflages itself within an environment of twigs and foliage, saving itself from predators by 'playing dead'.[81] The body is made continuous with its environment so as to be indistinguishable from it, but by extension his example of animal mimicry can be equated with an experience of 'radical desubjectivization'.[82] Space envelopes the individual, producing fear and a loss of agency. Caillois relates this to Eugène Minkowski's analysis of the experience of darkness as one in which the subject is enclosed or penetrated by the dark with boundaries between self and world dissolved. Just as the involuntary and automatic response of an animal's skin changes involuntarily and automatically, so human creativity might be thought to exist without a guiding control or will. Towards the end of Caillois's essay, he alludes to Gustave Flaubert's *The Temptation of Saint Anthony* as a 'general spectacle of mimicry' in which rocks resemble brains or plants look like animals. The saint is caught in the spell of the surrounding environment, desiring to 'penetrate each atom, to descend to the bottom of matter.'[83] The saint's strange and pathological attraction to the heart of matter is a strange *mise en masse* that reveals the radical erasure of the self.

Of course, 'playing dead' is the foundation of Hitchcock's *Vertigo*, the film itself an exploration of ego loss and death. Less explicit comparison has been drawn between Hitchcock's *Vertigo* and Smithson's *Spiral Jetty*, made just some twelve years later. We know that Chris Marker's films were a profound influence on Smithson and that the French director had in turn

80   Roger Caillois, 'Mimicry and Legendary Psychaesthenia', *October*, 31 (Winter 1984), 16–32.

81   Rosalind Krauss, 'Entropy', *Formless: A User's Guide*, ed. Yve Alain-Bois and Rosalind Krauss (New York: Zone Books, 1999), 78.

82   Krauss, *Formless*, 78.

83   Caillois, 'Mimicry', 31.

mediated his own obsession with Hitchcock's *Vertigo*, in *La Jetée* (1962). J. Hoberman reveals how indelible an influence *Vertigo* was on Marker in its strange magic of time travel and desire. Marker's imaginary narrator in *Sans Soleil* (1977) says, 'He wrote me that only one film had been capable of portraying impossible memory … In the spiral of [its] titles, he saw time covering a field ever wider as it moved away, a cyclone whose present moment contains, motionless, the eye.'[84] Smithson, Hitchcock, and Marker all approach the idea of leaping into the past through time's spiral configured on a cinematic reel.

Ostensibly, Hitchcock's *Vertigo* is a film about one man's consuming obsession with a dead woman or, rather, a living woman, who initially plays dead and then camouflages herself as the image of the dead woman. The protagonist, 'Scottie', played by James Stewart suffers from acrophobia after seeing his police colleague plunge to his death, however, the condition of vertigo is not just centred on his fear of heights, but on a dizziness of misrecognition in which living and dead, Judy and Madeleine become indistinguishable. His overwhelming desire for Madeleine means that after her 'death', he pressurizes Judy to change her clothes and hairstyle in order to resurrect the dead woman. This transformation is completed in a motel bedroom, flooded with a turquoise-green light, when Judy completes her metamorphosis into Madeleine by tying up her blonde hair into a twisted chignon, somehow echoed in the Stan Douglas's photograph, *Hair, 1948* (2010), discussed in Chapter 2. Hitchcock employs the green neon light and soft focus dissolves to produce this magical effect merging reality and illusion.[85]

84   J. Hoberman, 'The Lost Futures of Chris Marker', *New York Review of Books* (23 August 2012) <http://www.nybooks.com/daily/2012/08/23/lost-futures-chris-marker/>, accessed 7 February 2017.

85   American artist David Reed invoked this scene in his installation, *Judy's Bedroom* (1992) in which a loop of the transformation scene in the motel is played on a monitor within a bedroom installation suffused with the same turquoise light. Reed replicates the formulaic flower painting above the motel bed with his own swirling abstract placed above the bed in the installation, reinserting his painting into the film loop.

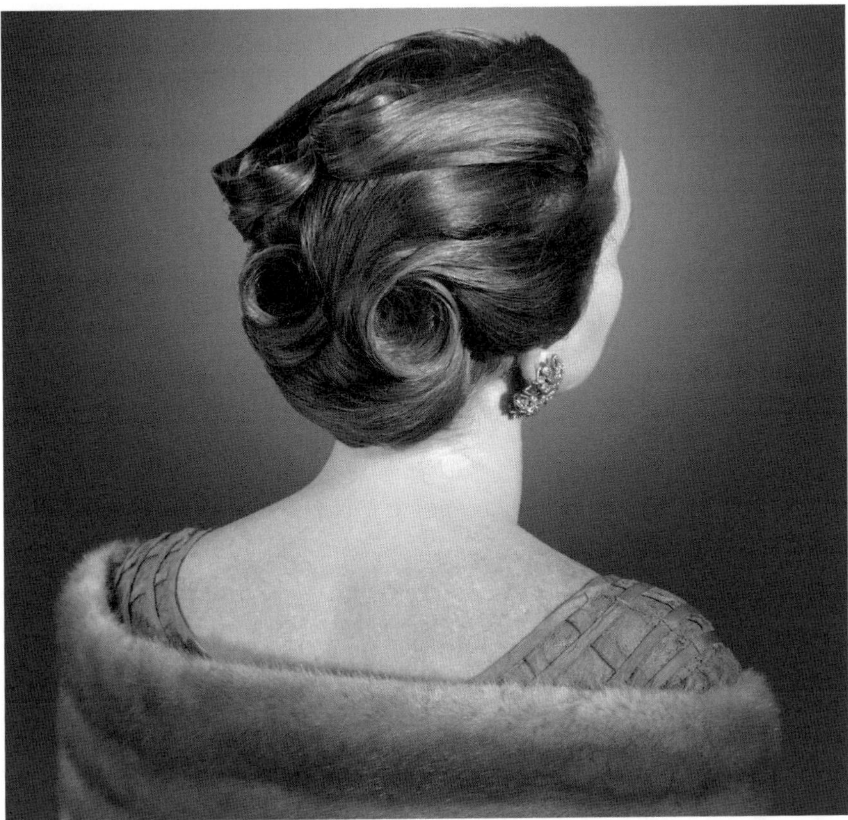

Figure 23. Stan Douglas, *Hair, 1948*, 2010.

Within the film scene, Scottie's re-connection to his dead lover is completed in a slow 360-degree kiss, in which they turn against a background
that morphs into the stables where they last met, finally dissolving into the
neon green light. Shooting this scene, Hitchcock placed James Stewart and
Kim Novak on a mechanical revolving platform with the changing backdrop created through projected transparencies.[86] However, the motif of
spinning is most evident in the opening title sequence by graphic designer,

86   Dan Auiler, *Vertigo: The Making of a Hitchcock Classic* (London: St Martin's Press,
       1998), 119.

Saul Bass, with hypnotic spiral forms emerging from the depths of Novak's eye. Bass collaborated with the computer animator, John Whitney, who reproduced *Lissajous* spirals by creating an animation stand that rotated continuously through the action of a floating pendulum. Whitney later worked with his brother, James to create a number of psychedelic animations based on the mandala (*Lapis*, 1977). Hitchcock represents Scottie's mental disintegration through an animation sequence in which a cartoon flower posy unbinds itself. Vivid Technicolor intensifies the hallucinatory quality of Scottie's nightmare with his face drowned in alternating colour filters, ending in his confrontation with an open grave and the image of falling. According to the artist and writer David Batchelor, falling is often carried in the metaphor of colour:

> Once again, the drug of colour begins to weigh heavily on our eyes; we become drowsy; we begin to lose consciousness as we fall under its narcotic spell; we lose focus; we lose ourselves in colour as colour frees itself from the grip of objects and floods over our scrambled sense; we drown in the sexual heat of colour ... And the Technicolor dream continues.[87]

Vertigo is a powerful metaphor; the partial or continual fall is that conscious revelation of death in life, a painful recognition of our ultimate disintegration into earth. Scottie's uncanny attraction to a dead woman and his fall forward into the grave within the dream is essentially his desire for self-annihilation.

It is difficult to ignore those compelling relationships between Smithson's film of *Spiral Jetty* and Hitchcock's *Vertigo*. Both artist and director were keen on the tour or tourist site. Smithson travelled extensively in the Yucatan, New Jersey, Utah, and abroad. Hitchcock's *Vertigo* is an inventory of San Francisco's vertiginous terrain of bridges and towers, just as Mount Rushmore was used for the final cliff-hanger in *North by Northwest*. For both artist and director, the spiral motif is fundamental within their interpretations of death and time. Time's spiral is even present in the scene set amidst the ancient Californian sequoias, when Madeleine

87   David Batchelor, *Chromophobia* (London: Reaktion, 2000), 43.

and Scottie examine the cross-section of a felled tree and she locates her birth among its concentric rings. For Smithson, time is paused and dilated through gravity's effects, and *Spiral Jetty* is the static cyclone that equates to the unravelling of film's frenetic rotations.

Let us finish with reference to *Vertigo* in a 1967 essay by Smithson, in which he detects the Gnostic sensibility at work in Hitchcock:

> The 'amateur craftsmanship' is everywhere in Hitchcock's *mise en scene*, from the ghastly green glow over roof tops in *Vertigo* to the diabolical sky in *The Birds*. Hitchcock's humor informs every terrible situation he takes his 'bad' actors through. His settings are a vast simulacra built by an evil demiurge, and peopled with frozen automatons.[88]

Technicolor's artificial glow is, in reality, an airless prison capable of inducing an attack of nausea. According to Batchelor, Technicolor becomes a metaphor for falling.[89] What Smithson recognizes in Hitchcock's cinema is a dichotomy of knowledge and oblivion. In Smithson's line of thinking, the cinema is essentially a jail best located in an old mine or cavern, where the flickering shades briefly illuminate the world of gross, heavy matter that confines us. As explored in Chapter 2, Manichaeism is a strand of heretical Gnosticism, based on the dualistic principle of light and dark, illumination standing for the idea of knowledge. In Smithson's essay, the moviegoer is described as a 'hermit' dwelling in the 'flickering shadows' and experiencing 'blurs of many shades'. Operating between light and darkness, driven by a spiral mechanics and peddling mere illusion, film exemplifies Gnostic doubt. By giving ourselves up to film illusion, we also abnegate responsibility onto cinema in the manner of Kubrick's HAL, a 'deafening pale abstraction controlled by computers.'[90] Returning to the early Technicolor fantasy, *The Wizard of Oz* again, we arrive at the Emerald City, Smithson's 'crystalline build-up'. Toto the dog follows a scent and tears back the lurid green curtain to reveal an old man moving levers and cranks to sustain the spectacular illusion. His 'demiurge' character cannot

88   Smithson, *Collected Writings*, 353.
89   Batchelor, *Chromophobia*, 41.
90   Smithson, *Collected Writings*, 139.

have been lost on the artist. Following Toto the dog, Smithson challenges the idea of film as a medium that normally animates and preserves life, instead pricking its illusion to reveal the blind impersonality at the heart of its crystal maze. Gravity is a prison-house, emblematic of Gnostic doubt and blindly sadistic to human vanity, agency, and aspiration.

CHAPTER 5

Critical Mass

> If my rage at the impoverishment of ideas, narcissism, and disguised sexual exhibitionism of most dancing can be considered puritan moralizing, it is also true that I love the body – its actual weight, mass, and unenhanced physicality.[1]

My exploration into the role of gravity in contemporary art has focused on a number of areas: I have questioned how contemporary artists have responded to architecture and the space of the fall; how they play with gravity in terms of the agency lodged in animate and inanimate matter; and how they have made work in response to the history, spaces, and labour of heavy industry. In the previous chapter, I have explored how gravity relates to vertigo and ideas of industrialized entertainment as configured through Robert Smithson's interest in gravity and vertigo. The essential modality of how we seize gravity is captured in states of falling. This final chapter takes account of how gravity has been invoked through the body in forms of dance, sculpture, and performance against a narrative of agency reduction and enlarged possibilities for participation. In contrast to ballet's light flourishes, dance takes on mass and momentum with new variations of uninflected movement in space, dictated by gravity's imperative more than its illusionistic sublimation. This submission to gravity is matched in the way sculpture moves off the plinth and acquires a softness that reminds us uncannily of the body. The forms of manufacture and industry discussed in Chapters 2 and 3 re-emerge as fallen matter or junk. Chicken wire, old mattresses, cardboard, burlap, and stockings are used in sculpture, their lumpen quality suggesting a Gnostic image of artist as demiurge.

1   Yvonne Rainer, 'The Mind is a Muscle' statement (1968) in Catherine Wood, *Yvonne Rainer: The Mind is a Muscle* (London: Afterall Books, 2007), 41.

In 1947, just a few years before my account begins, Jackson Pollock painted a series of 'cosmic' works inspired by the night sky with titles such as *Galaxy, Comet, Lucifer*, and *Reflections of the Big Dipper*. These works formed part of his transition to a painting method, characterized by dripping and pouring paint directly onto a large canvas placed on the ground. Some of his huge works sagged under their own weight, heavy with canvas and paint layers. However Pollock's new method in painting became more than the sum of its parts; not just enlarged painterly gestures, but more consciously felt arbitrations of scale, space and force. He extended his physical agency with tools and the viscosity of paint to stretch the frame or field of conventional painting. Gravity was the footbridge that connected his body's action under its own mass and the entropic forces poised to engulf painting and frame. In his now famous essay, 'The Legacy of Jackson Pollock', Allan Kaprow describes how we need to be 'acrobats' to grasp the energy-dialectic of his work, shuttling between both centripetal and centrifugal forces.[2] By experimenting with different tools and the viscosity of paint, Pollock set up the conditions under which his body and the force of gravity would perform together, moving his weight into the trajectory of the paint.[3]

Whilst my account of gravity in art might naturally encompass Pollock's dripped paintings, his work is actually understood within a history that still validates paintings as responses to our 'vertical' orientation and existence in nature. In *Other Criteria*, Leo Steinberg argues that the real shift from vertical to horizontal plane occurs in the 1950s first with Robert Rauschenberg and is marked by a content shift from nature to culture.[4] (Pollock's paintings still belong to this vertical orientation.) Steinberg claims of the artist:

2   Allan Kaprow, *Essays on the Blurring of Art and Life*, ed. Jeff Kelley (Berkeley and Los Angeles: University of California Press, 1993), 5.

3   Paul McCarthy lampoons the abstract expressionists' emphasis on creative subjectivity in *The Painter* (1995), in which he wears a bulbous prosthetic nose and exaggerated rubber hands, whilst playing with paint and other foodstuffs such as mayonnaise and ketchup. In one scene, he drags his entire body through a paint-strewn floor.

4   Leo Steinberg, *Other Criteria in Twentieth-Century Art* (Oxford: Oxford University Press, 1976), 54.

> He lived with the painting in its uprighted state, as with a world confronting his
> human posture. It is in this sense, I think, that the Abstract Expressionists were still
> nature painters. Pollock's drip paintings cannot escape being read as thickets; Louis'
> *Veils* acknowledge the same gravitational force to which our being in nature is subject.[5]

Steinberg uses the printing term, 'flatbed picture plane' to anoint a key change of address from vertical to horizontal. As a result, Rauschenberg's images and objects tacked down to a surface needed a picture plane that was as 'hard and tolerant as a workbench' and resembled a 'disordered desk or an unswept floor'.[6] Steinberg's concept of the flatbed picture plane is characterized by images that do not first contain an optical idea or impression, but might, as in Rauschenberg's case, contain a work surface that operates as a 'dump' or as a repository for the 'detritus of communication' as ideas issue from his process.[7] He even took the horizontal plane of his own bed (*Bed*, 1955), together with quilt and pillow, dripped oils over it and mounted it like a painting.

Performance is, however, central to the many interpretative negotiations of Pollock. Hans Namuth's photographs and film captured his dripping dance in process and performance. In his colour film of 1950, he shows Pollock performing a hunched, foot weaving side step across the length of a canvas placed on the ground. In another scene, Namuth places the camera underneath a glass with the artist dripping skeins of paint across the surface from above. These recordings of Pollock's artistic process have spawned many imitations and parodies within performance ever since. Amelia Jones's reading of what she terms the 'Pollockian performative' takes account of a range of body practices that were constructively shaped by Pollock's conjugation of body and paint from Kazuo Shiraga's solitary mud wrestling (*Challenging Mud*, 1955) through Yves Klein's *Anthropometries* to Shigeko Kubota's *Vagina Painting* of 1965. By harnessing the force of gravity, artists and performers could free their own subjective stranglehold on form making to reduce agency, as discussed in Chapter 2.

5   Steinberg, *Other Criteria*, 84.
6   Steinberg, *Other Criteria*, 88.
7   Steinberg, *Other Criteria*, 88.

From the 1960s onwards, performers and artists based in New York City, such as Robert Morris, Yvonne Rainer, Claes Oldenburg, Trisha Brown, Steve Paxton and Simone Forti found themselves working in empathetic conjunction. Jim Dine, Allan Kaprow, and Oldenburg staged happenings and constructed environments in Greenwich Village, whilst dancers, Rainer, Paxton, Brown and Morris (amongst others) presented choreographies at the Judson Dance Theatre in Greenwich Village, organized by Robert Dunn. These dance experiments stemmed from the workshops organized by Paxton, Dunn, and Rainer from 1962, prior to early performances in January 1963. In May 1965, the young choreographer, Steve Paxton organized the *First New York Theater Rally* with revivals and new dances by Robert Morris, Yvonne Rainer, and Trisha Brown, as well as 'Happenings' and performances by Öyvind Fahlström, Robert Whitman, and Claes Oldenburg.[8] The year 1965 also marked the high point of performance and experimentation at the Judson Dance Theatre. However, other strands of influence flowed into Judson, rooted in the experimentation that had taken place at Black Mountain from 1944 onwards between artists, musicians and dancers, notably Rauschenberg, John Cage and Merce Cunningham.

Cunningham set up his own dance company in 1953, sharing Cage's interest in chance and exploring the body's quotidian routines and gestures. This was to shape the direction of modern dance with its borders made porous with sculpture, environment, and everyday gesture. The dancer's interest in chance effects moved dance into an activity freed of overt subjectivity and professional ownership. Importantly, Cunningham articulates modern dance as reaction to ballet's general appeal to lightness with its climax of lifts and pirouettes *en pointe*. Instead, he invokes gravity as it draws the body down, in order to reflect physical life more tangibly, noting: 'one of the best discoveries the modern dance has made use of is the gravity of the body in weight, that is, as opposite from denying (and thus affirming) gravity by ascent into the air, the weight of the body in going with

---

8   Sally Banes, *Terpsichore in Sneakers: Post-Modern Dance* (Middletown: Wesleyan University Press, 1987), 14.

gravity, down.'[9] From Cunningham's example, choreographers broadened the meaning of dance to include everyday rituals and quotidian gestures, taking dance into unconventional settings such as a dump container, a church, rooftops, or the street. Somehow the derelict and fallen condition of New York City worked its way into sculptures and performances with gravity more consciously bidden into objects and bodies.

Junk and detritus became art critical elements that seemed to augur the idea of performance as throwaway. As Kaprow claims: 'The use of debris, on another level, reassured an art world then occupied with exploring the junk of our throwaway culture – *junk* was a password – and not only was the act of sweeping easy, but it was also a nonart act, disposable like its material.'[10] This was accompanied by a deliberate effacement of personality, since the value of a person's interior life or deep psychology was downgraded. The body and junk somehow echo each other in this throwaway aesthetic, just as Beckett's reading of Geulincx led to the idea that the body was worth nothing.

In his body-inflected sculptures, Oldenburg collects and subverts Pollock's use of scale and dripped paint in his oversized floppy domestic objects.[11] Oldenburg's *The Street* (1960) contains sculptures constructed out of the containers of street garbage, cardboard boxes, and burlap, laid over with crude graffiti-like painting. These metamorphic objects and fragmented figures in *The Street* are suspended from the wall, moving with the air to echo a shifting scene of people walking down the rough streets around Lower East Side. Originally shown as part of a two-person exhibition, *Ray Gun Show* with Jim Dine, *The Street* opened at the Judson Gallery (Judson Memorial Show) in January 1960, animated by Oldenburg in a series of performances located in the grotty basement of the church. In *Snapshots of a City*, Oldenburg and Patty Mucha (later his wife) wore old bandage-like rags and writhed around in the street junk they had accu-

---

9   Merce Cunningham, *Merce Cunningham Dancing in Space and Time*, ed. Richard Kostelanetz (New York: Da Capo Press, Inc., 1998), 38.

10   Allan Kaprow, *Essays on the Blurring of Art and Life*, 184.

11   Joseph Kozloff, *Renderings: Critical Essays on a Century of Modern Art* (London: Studio Vista Ltd, 1970), 223–35.

mulated for *The Street*. As Susan Sontag noted of the materials used in the messy 'Happenings', one could not distinguish among people and objects or clothing and costume because 'people were often made to look like objects by enclosing them in burlap sacks, elaborate paper wrappings, shrouds and masks'.[12] By wearing and performing the junk he had scavenged from the street, Oldenburg reanimated the forgotten vitality of dead, discarded stuff. This anthropomorphism is evident in certain works by Bruce Conner, such as *RATBASTARD* (1958), in which he cut open a canvas, stuffed and bandaged it with nylon stockings so that it looked as if 'its innards were coming out'.[13] By stuffing rucksacks and canvases, Conner's objects look as if they are brimming with thick matter, replete with fleshly wadding that is barely contained by the skin of the nylon tights.

Sally Banes characterizes the choreography at the Judson as 'democratic' with dancers finding new directions by erasing subjectivity, authority, and *telos* from dance's accepted hierarchies and structures. Dance became a site of critique that foregrounded changing degrees of connection between gravity and willed effort, but it also heralded a closer space of association between dancer and audience with no stage or special costume. The figure of the fall crept into the rhetoric of dance and, with it, the indeterminacy of accidental and incidental movements. Performers such as Yvonne Rainer, Steve Paxton, and others of the Judson Dance Theatre probed the body's natural physics, examining its normative effects across a range of physical habits and routines under conditions of force, spatial environment, and human will. By exploring non-heroic movements, dancers and choreographers invoked an object-hood or estrangement of the body. These radical experiments meant taking dance into non-traditional settings: the tops of buildings, the streets of Greenwich Village, a dumpster, and even a church. Allan Kaprow formulated the notion of 'Everywhere as playground' as an antidote to the tired puritan ethic of work as formalized routine.[14]

---

12   Susan Sontag in *Art Since 1900: Modernism, Antimodernism, Postmodernism*, ed. Hal Foster, and others (London: Thames and Hudson, 2004), 453.

13   Anna Dezeuze, *Almost Nothing: Observations on Precarious Practices in Contemporary Art* (Manchester: Manchester University Press, 2017), 58.

14   Kaprow, *Essays on the Blurring of Art and Life*, 112.

In Judson dance, pedestrian movements are more significant than displays of agility and gracefulness. Incorporating everyday movement into dance formed part of the general breakdown between art and life so that the body was moved in a way that indicated spontaneous creation. This dissolved notions of the artwork as fixed and finished commodity. As a result, art could happen anywhere and performance was no longer tied to a proscenium space of illusion or geared towards a transcendental high. Dancers of the Judson school were focused on the phenomenology of gravity-ground embodiment, reflecting on how everyday movements, tasks and actions might leak into the 'medium' of dance. Certain machine or task-like functions were ascribed to the performer so that labour, work, and performance were conflated and the body was explored as de-subjectivized mass or as estranged object. Associations between human will and personal expression were rejected. Instead, dancers interrogated shared conditions of everyday experience or sensation such as falling, walking, and lying down with process privileged over performance.

Evolving out of experimentation at the Judson Dance Theatre, *Contact Improvisation* was a dance technique that was developed around 1970 by Steve Paxton, Nancy Stark Smith, and Trisha Brown amongst others. It involved a minute appreciation of the body's momentum-in-mass, feeling its relation to gravity and to weight exchanges with other bodies. Simple tests of transferring the body's weight to a partner and creating new movements by letting go of control were ways of feeling the body's downward pull or its tipping point.

In Trisha Brown's practice, gravity is a common bond, from which to examine different embodiments of space. Her *Leaning Duets*, a version of which was filmed on Wooster Street, New York, in 1970, are a set of physical duets in which each pair walks as a single unit, their bodies connected by the hand in a single act of walking. These simple pairings are reminiscent of the dyadic counterpoint developed between Wood and Harrison in *Twenty-Six (Drawing and Falling Things)*. Counterbalanced, the performers lean out from a single centre of gravity created at the join between their feet. They walk until one of them exerts greater drag and upsets the balance. These leaning works also recall the bends and inclinations of Gilbert and George or Bas Jan Ader with emotional connotations of togetherness in co-participation.

Let us recap Erwin Staus's definition of 'inclination' as a 'reaching out': 'Inclination first brings us closer to another. Inclination, just like leaning, means literally 'bending out' from the austere vertical.'[15] In Staus's reading of uprightness, it is the vertical condition that marks human beings as aloof and cold. Yielding to the body's weight, resting, lying down, sinking to the ground, these leaning dances exude warmer connotations of humility and our common bond in gravity. In Brown's choreographies, intensity is located in the diagonal, in the body tips and tilts geared at testing gravity. As Martin Hargreaves summarizes, 'Brown used gravity to twist an idle perambulation into an act of full-bodied performance.'[16]

Contact improvisation involves a free play of movement guided by the body's natural momentum. In releasing muscular tension, the body flops, droops, and falls in harness to its weight or against the momentum of another dancer. In 1987, Steve Paxton reviewed the development of contact improvisation in a 23-minute video, *Fall After Newton*. His explanation for this type of dance runs as follows:

> When an apple fell on his head, Isaac Newton was inspired to describe his three laws of motion.
>
> These became the foundation of our ideas about physics. Being essentially Objective, Newton ignored what it feels like to be the apple.
>
> When we get our mass in motion, we rise above the constant call of gravity toward the swinging, circling invitation of centrifugal force. Dancers ride and play these forces.[17]

Paxton was to recognize how the energy given up to resisting gravity is just taken for granted. Contact improvisers learn how to give into the body, releasing control and producing spontaneous movements out of the body's weight and momentum. Dancers introduce risk and chance by

15   Erwin Staus, 'The Upright Posture', *Psychiatric Quarterly*, 26 (1952), 529–61 (539).
16   Martin Hargreaves in *Move: Choreographing You*, ed. Stephanie Rosenthal (London: Hayward Publishing, 2011), 67.
17   Steve Paxton in *Dance: Documents of Contemporary Art*, ed. André Lepecki (London: Whitechapel Gallery and Cambridge, MA: The MIT Press, 2012), 62.

inclining the body, leaning, or allowing the body's weight to be absorbed slowly by sliding into the pathway of someone else. His explanation in *Fall After Newton* reframes gravity within ideas of agency beyond subjective control, imagining the feelings of the falling apple. For Paxton, radical defamiliarization occurs as bodies combine in movement, detouring from their normal alignment with gravity and allowing a natural momentum to lead the dance like the paint that journeyed downwards from Pollock's dripping dance.

Mass, Muscle and Movement

In 1968, Yvonne Rainer created a multi-part performance with the unusual title of *The Mind is a Muscle*. What relationship between the body's brain and movement did Yvonne Rainer envisage in terms of the choreography? How is the mind a muscle? Certainly, we think of muscle as our material core; weight-bearing, common, and impersonal tissue. Rainer perhaps intimates that the mind is better grasped in its embodiment, grounded rather than transcendent. Stripping away the gestural traditions of dance, Rainer's evening-long performance, *The Mind is a Muscle* listed the following short works: *Trio A, Trio B, Mat, Stairs, Act, Trio A1, Horses, Film*, and *Lecture* with an intermission after *Act*. In its original showing, the dances involved seven performers, including Barbara Lloyd and Steve Paxton, as well as Rainer herself. In Scene 4 (*Act*), she includes two trapeze swings and a magician in a black suit and top hat (Harry de Dio) performing juggling and magic tricks to the side, a conceit of illusionism that the dance rejects. A variety of non-spectacular actions and movements make up the different sections, with people carrying out simple tasks, actions or work-like motions: lifting 12 lb dumbbells, climbing up and down ladders, walking, lifting one another up, moving like a herd of animals, walking with arms swung like rocks suspended on a string. Rainer's overall instruction to the performers is to exhibit: 'Task- or work-like movement with a "factual" quality, in part presented with a smooth, apparently even, energy continuum. Eyes averted from engagement with the audience to create the effect of a blank gaze into the mid-ground.'[18] No climaxes or opportunities for personal expression are permitted. Instead, the performers have to concentrate on producing an even flow of energy without crescendo or dramatic high point.

In place of phrasing or narrative flow, the dancers roll and heave their bodies, scattering the body's centre across limbs and props in alien movements. Rainer stresses a conjunction with the real, localized weight of the body as its natural lag hits the ground. The body's drag extends also to facial expression. Rainer fashioned an expressionless face, known as the 'Judson stone face' to neutralize hierarchies and remove personality.[19] The dance evokes the body's literalism as muscle and impersonal matter. As Catherine Wood writes: 'Rainer speaks about her body in a decidedly laconic and anti-metaphysical way as bare matter. Her pragmatist stance led to an aesthetic of radical laconicism that bypassed the fundamental question of bodily 'authenticity'.[20] Choreographed earlier in 1966, *Trio A* is perhaps the best-known element of *The Mind is a Muscle*, re-performed by Rainer and filmed in 1978. This short routine of uninflected movements is utterly laconic as if the goal were to suppress any sense of hierarchy or stylized gesture. The normal co-penetrations of gravity and grace found in dance, acrobatics, or sport are curbed so that any instinct or natural latency of the body is arrested. Even short repetitions of step movements are integrated without any notable patterns emerging; a set of hand sways and leg lifts like the slow ghost of a 'Charleston' dissolve immediately into other body motions. In moments where the body is given to exert itself in a small set of jumps, the energy expenditure is highly controlled, eliminating any sense of climax. The way in which Rainer frames her body as impersonal matter is politically radical and critical. In her book, *Vibrant Matter*, Jane Bennett contends that an approach to matter based on vital materialism captures a commonality between body and non-organic life. This is subtended by a radical approach to thinking about materialism, and by extension, consumerism. She writes: 'Vital materiality better captures an "alien" quality of our own flesh, and in doing so reminds humans of the very radical character of the (fractious)

18   Wood, *Yvonne Rainer*, 7.
19   Wood, *Yvonne Rainer*, 28.
20   Wood, *Yvonne Rainer*, 36.

kinship between the human and the non-human.'[21] In fact, as she argues, our bodies are not entirely human: there are parasites and microbiomes nestled in or on the body. She therefore sees the political potential in this view of the body: 'If more people marked this fact more of the time, if we were more attentive to the indispensable foreignness that we are, would we continue to produce and consume in the same violently reckless ways?'[22] Such radical estrangement from our own bodies is caught in the dramatic interval of the fall in moving from subjective body to falling object.

Opening *Trio A*, Rainer bends her knees slightly and swings her arms in a vertical drop against her body, clockwise and anti-clockwise. A few moments later, she holds her arms out horizontally at 90 degrees to the body and rotates them. In her analysis of Yvonne Rainer's *The Mind is a Muscle*, Catherine Wood notes a contradiction between 'presentations of the body-as-object, either as heavy weight to be lifted or as mannequin-style automaton.'[23] This mannequin style of presentation subtracts from the idea of dance as leading from a type of soul-centred expressivity. It was entirely this style of expressivity that nineteenth-century writer Heinrich von Kleist had critiqued both theologically and aesthetically in his dialogic essay, *On the Puppet Theatre*, published in 1810.[24]

Puppet Envy

Kleist's curious treatise is a remarkable augury of modern ideas of 'gravity and grace' in philosophy and performance. His vision of the puppet theatre anticipates a different contract with grace in respect of nature, the machine and the soul. In this short essay, the narrator and Herr C., a professional

21   Jane Bennett, *Vibrant Matter: A Political Ecology of Things* (Durham: Duke University Press, 2010), 112.

22   Bennett, *Vibrant Matter*, 113.

23   Wood, *Yvonne Rainer*, 27–8.

24   Heinrich von Kleist, 'On the Puppet Theatre' in *Selected Writings*, ed. and tr. David Constantine (Indianapolis: Hackett Publishing Company Inc., 2004), 411–16.

dancer, debate a radically different quality and meaning of physical expression. Puppets, according to Herr C., objects constructed out of wood and cloth, are free because they lack awareness of what it means to inhabit a real body. Their movements are graceful because they are controlled through wire and string by a puppet-master above, the master remaining static while the puppet dances. He explains how the puppet is:

> incapable of affectation. – For affectation occurs, as you know, whenever the soul … is situated in a place other than a movement's centre of gravity. Since the puppeteer, handling the wire or the string, can have no point except that one under his control, all the other limbs are what they should be: dead, mere pendula, and simply obey the law of gravity; an excellent attribute that you will look for in vain among the majority of our dancers … these puppets have the advantage of being resistant to gravity.[25]

Puppets do not suffer the fatigue or the instinct for collapse experienced by human dancers. By contrast, Herr C. exemplifies lack of grace as a characteristic found at the ballet. When Mr F. dances the role of Paris and offers an apple to Venus, his movements are affected and unnatural. Unlike the marionette, Mr F.'s soul is displaced from the body's centre of gravity so that 'his soul – it is painful to see – is actually in his elbow'.[26] In a 'Bernini like' performance, the personally expressive line is moved self-consciously out of the centre of the body, out of kilter with gravity and the creator-puppeteer. Helmut Schneider explains: 'Kleist's essay turns the value hierarchy of the classical discourse [on grace] upside down, or, more exactly, inside out. It questions the priority of the interior over the exterior, the spiritual over the corporeal, the meaningful over the accidental, the metaphor over the literal, the signified over the signifier, the soul or mind over the body'.[27] As a result, gravity is the necessary corrective to this traditional model of grace.

Herr C. goes on to extol the true grace found in the dances of cripples with prosthetic limbs made by English craftsmen and reflex feints of a fencing bear, who outwits the fencing master. Kleist's suggests that it would

25   Kleist, *Selected Writings*, 413.
26   Kleist, *Selected Writings*, 413.
27   Helmut J. Schneider in Nicholas Ridout, *Stage Fright, Animals, and Other Theatrical Problems* (Cambridge: Cambridge University Press, 2006), 25.

be better to replace the human performer with a machine. A marionette would do the trick because it would mean avoiding the shameful meeting of eyes between performer and audience within any theatrical encounter.[28] Philosophically, grace (that strange fusion of the ethical and the physical) is turned through another twist of irony from E. T. A. Hoffman's dark reading of the mechanical double (*The Sandman*) to the enviable condition of being a puppet. As John Gray confirms in his reading of Kleist's thesis:

> How could a puppet – mechanical device without any trace of conscious awareness – be freer than a human being? Is it not this very awareness that marks us off from the rest of the world and enables us to choose our own path in life? Yet as Kleist pictures it, the automatism of the puppet is far from being a condition of slavery. Compared with that of humans, the life of the marionette looks more like an enviable state of freedom.[29]

However unconsciously, dance in post-1960s America seems to refract Kleist's strange vision of grace and the body's relationship to gravity. Merce Cunningham and Yvonne Rainer's dance vocabulary, involving pedestrian or everyday movement, corresponds with Kleist's emphasis on embodiment as a kind of unvarnished truth or unmediated experience. Herr C. rejects Bernini style expressivity found at the ballet. Similarly, Rainer assumes the deadpan ('Judson stoneface') and favours an idea of the body that is seen as an object to be moved and handled. Individuality and authorship are subservient to mechanical efficiency.

There are other important parallels. Kleist's typical baroque dancer suffers from affectations that arise from the postlapsarian condition, robbing him of natural grace. Herr C. explains to his interlocutor how a young dancer had become so self-consciously aware of his own grace and beauty that 'an invisible and incomprehensible power seemed to settle like a iron net over the free play of his manners'.[30] Henceforth, the performer could only be regarded as a fake stand-in, performing a charade of weak illusion

28   Ridout, *Stage Fright*, 25.

29   John Gray, *The Soul of the Marionette: A Short Enquiry into Human Freedom* (London: Penguin, 2016), 5.

30   Kleist, *Selected Writings*, 415.

and possessing consciousness but no grace. This echoes Rainer's insistence that 'narcissism' and 'exhibitionism' were to be banished from dance. Studio mirrors were not used at Judson in order to prevent any 'preening' feed-back.[31] In Kleist's other example of grace, the fencing bear is characterized by movement that arises out of pure reflex and unselfconsciousness, exemplary of the purely organic world. His essay concludes: 'Grace will be most purely present in the human frame that has either has no consciousness or an infinite amount of it, which is to say in a marionette or in a god.'[32]

John Gray's contemporary reading of Kleist's puppet treatise draws out its inherent Gnosticism within a vision of the world as fleshly error and obdurate matter. The marionette is operated mechanically as in the automatic turning of a barrel organ, but it is superior to human beings who exist in the half-light of cosmic error. Gray writes: 'When Herr C. tells the narrator that he should read the third chapter of Genesis, Kleist points towards the most radical of these traditions – the religion of Gnosticism.'[33] He goes on to argue that Kleist's text is a radical assertion in which: 'Real freedom would be a condition in which they would no longer labour under the burden of choice – a condition that could be attained only by exiting from the natural world.'[34] If Kleist's human performer represents degraded illusion, then this is framed within Gnosticism's dark reading of humanity and matter as error.

Alienation is the keynote of these readings of the puppet and yet Kleist reverses classical ideas of grace in his reading of the puppet and dummy, reflecting critically on the abstract value of the machine-like. Control is relinquished in the mindless 'turning of a handle to play a barrel-organ.'[35] There is always an ambiguity of effect and meaning within the concept of automatism: on the one hand, the automatic signals chance effects so as to free up the individual's unconscious (not as evident in Surrealist practices as one might think). On the other hand, automatism negates human volition

31   Wood, *Yvonne Rainer*, 33.
32   Kleist, *Selected Writings*, 416.
33   Gray, *The Soul of the Marionette*, 8.
34   Gray, *The Soul of the Marionette*, 8–9.
35   Kleist, *Selected Writings*, 412.

or thought, promoting the machine-like over personal expressivity. The machine lacks interior thought or consciousness. Is the work of controlling puppets by turning a handle therefore a more satisfactory pleasure? Herr C. seems to suggest that transferring the dance of puppets 'wholly into the realm of mechanical forces' is a desirable option.[36] This was paralleled in dance practice in the use of mechanical movements of routine. Whatever the answer, Kleist exposes bourgeois aesthetic ideology as no more autonomous than the cheap charade of slave-master routines in the puppet theatre. These spectacles in fact provide a more authentic vision of grace without affectation.[37]

Kleist's aesthetics of performance is found within the low-brow puppet theatre, in the burlesques positioned at the market-place, as opposed to the slick operation of the provincial theatre. Is the tawdry commercialism of the puppet theatre, like the circus, a more transparent embodiment of labour transaction between *artiste* and spectator? Entertainment and leisure industries have diverted our gaze onto performers who do not answer back. Circus performers use tendons and muscles to amaze us, but in turn, we enter a temporary fog, surrendering the potential thrust and stretch of our own lower limbs. Georges Seurat's painting of the circus (*Circus*, 1891), referred to in Chapter 3, was a darkly ironic reflection of this circumstance with the stupefied audience reflected back to the viewer across a vision of spectacular movement. What touched modern artists such as Picasso and Léger was the pathos associated with the circus *artiste* in a commodity exchange of performance for money. The circus performer, in contract and harness to the invisible force of gravity, belonged to a category of excluded outsider forced into painful exertions every night in order to survive. If we can replace the human performer with a machine, could this produce a more palatable entertainment, eliminating the embarrassment of labour relations?[38]

---

36   Kleist, *Selected Writings*, 412.

37   After Kleist's reverses the idealized concept of physical grace into the mechanical realm, the twentieth-century reading of grace is more closely associated with the awkward pedestrian movements of Charlie Chaplin.

38   Ridout, *Stage Fright*, 26.

The ramifications of Kleist's essay seem to resonate within this strand of dance and performance aesthetics in the 1960s. As Rainer states in her manifesto: 'No to spectacle no to virtuosity no to transformational and magic and make-believe no to the glamour and transcendency of the star image.'[39] She explores questions of bodily effort as equivalent to human labour in complex and ambiguous ways, but most importantly, anti-spectacular dance was a medium that resisted commodification and reification. As Catherine Wood states: 'Tautologically, the work (of art) is simultaneous with the work (effort) of its execution.'[40] Rainer's *Quasi Survey* of 1966 establishes equivalence between factory fabrication and 'energy equality and found movement'; unitary forms or modules are thus paired with a performance that maintains equivalence between parts. Repetition mimics the experience of factory fabrication and reflects the use-value of the human body, aesthetically transformed, back to the audience. For dancers like Rainer, everyday movements such as working or walking became a more fruitful terrain for creative exploration.

Simple pedestrian movement mixes a 'starting forth' into space with a momentary arrest on earth. It is continual negotiation or compromise between lateral and vertical states of the body. Walking is also the ultimate process art, a wayfaring rather than a goal, a physics of the body that evaporates at every step. It is an action that bends easily to phenomenological analysis. Solace is found in the thickened limit of earth and the act of walking gives artists a right of passage to expanded fields of investigation. This constitutes a shared terrain for both artists and dancers. As explored in my second chapter, Bruce Nauman's awkward pacing across a square perimeter in *Slow Angle Walk (Beckett Walk)* (1968) or Samuel Beckett's hunched and manic walkers in *Quad* (1986) request the reassurance of gravity to find an answering thud and pressure transmitted from earth to body. In the extended walks of Richard Long, Hamish Fulton, and Francis Alÿs, locomotion is a way of retreating to, or resisting the amnesic tread of routine.

39   Yvonne Rainer, 'On Dance for 10 People and 12 Mattresses Called Parts of Some Sextets' (1965) in Lepecki, *Dance*, 48.
40   Wood, *Yvonne Rainer*, 27.

Walking is a route to re-examining how habits and postures become so easily harnessed to regulation.

Nauman's *Device to Stand In* (1966) is a brass-wedge slot for spectators or 'performers' to stand in like an enlarged slipper, drawing their consciousness to the underneath and downwards of the ground, whilst collapsing the distance between sculpture and body. For Nauman, this object re-orientates its user to the ground, stripping out those habits of perception whereby the ground is a surface continually in retreat. Michel Serres overturns such bodily amnesia, reflecting on how 'everything comes from your base, positioning and posture on the earth, from your balance, from your instinctive gripping of the ground with the soles of your feet, from your grasping hold of long roots with your toes ...'[41] Gravity is the permanent undercurrent that ties us to everyday movement. By making small dramas out of its quotidian effects, performers have lifted these experiences into our awareness so that everyday movement creeps into the more open space of performance. Conversely, in vaudeville performance, the gentleman tramps, 'Flanagan and Allen' (referenced by Gilbert and George) walk onto the stage as if it were continuous with the street they apparently live on, furnished with benches and lamp posts. By including non-heroic movements such as walking in dance and performance, modern dance of this period named and enfolded other forms or patterns of movement within a broadened concept of choreography. As Susan Leigh Foster suggests: 'troop movements to board-room discussions, to dog training and computer flow', an enlarged meaning of choreography intimates 'that bodies are being choreographed by a very authoritarian, Manichean even, wielding of power.'[42] Ethnographic, psycho-geographical, or psychological approaches have made walking into an art that critiques how the body is corralled within space and how walking is a radical end in itself, antagonistic to the idea of the body as productively efficient and goal orientated.

41   Michel Serres, *The Five Senses: A Philosophy of Mingled Bodies*, tr. Margaret Sankey and Peter Cowley (London: Continuum International Publishing, 2008), 316.
42   Susan Leigh Foster in Rosenthal, *Move*, 37.

Graviphilia

In 1934, the Polish writer, Bruno Schulz wrote a short story in his collection, *The Street of Crocodiles*, which reflects some aspects of Kleist's *On the Puppet Theatre*, both being animated by a similar Gnostic spirit.[43] In Schulz's short chapter, 'A Treatise on Tailors' Dummies', the narrator's father sees value in the dead mannequin because it exhibits a malicious state of matter that is continuous with human beings: 'Can you imagine the pain, the dull imprisoned suffering, hewn into the matter of that dummy which does not know why it must be what it is, why it must remain in the forcibly imposed form which is no more than a parody.'[44] Lumpen matter is woven invisibly into forms by an evil demiurge so that in Gray's reading: 'individuality is a type of theatrical display, in which matter assumes a temporary role – a human, a cockroach – and moves on.'[45] As a result, the father argues, human beings are more interested in unlocking the seams and hidden forms of matter rather than being deceived by the seamless dream of life's surface. Moreover, humankind is interested in matter's 'creaking, its resistance, its clumsiness ... its inertia, its heavy effort, its bear-like awkwardness.'[46] As a result, the narrator's father favours humble things: 'we shall give priority to trash. We are simply entranced and enchanted by the cheapness, shabbiness, and inferiority of material.' He continues, 'the deep meaning of that weakness, that passion for colored tissue, for papier-mâché ...'[47] The father's ennobling of trash augurs the value placed on shabby materials in what Anna Dezeuze describes as the junk aesthetic of the 1950s and 1960s. Cardboard, kapok, burlap, old mattresses, and rags were the constituents of objects and performances made by Oldenburg, Conner and Rauschenberg.

Discarded objects degrade into polymorphous existences, their collapse and disintegration severing the reassuring connection between matter and

43   Gray, *The Soul of the Marionette*, 19.
44   Bruno Schulz, *The Street of Crocodiles* (New York: Penguin, 1977), 64.
45   Gray, *The Soul of the Marionette*, 24.
46   Schulz, *The Street of Crocodiles*, 62.
47   Schulz, *The Street of Crocodiles*, 62.

the named or recognizable object, 'grittily resisting incorporation into a smooth esthetic whole'.[48] Released from a particular incarnation, matter is free to take on other roles. Vital materialists are keen to resurrect matter from its designation as dead or defunct. As the father warns in Schulz's story, 'lifelessness is only a disguise behind which hide unknown forms of life'.[49] Within the context of post-war art, found objects, junk, and detritus were valorized and put into new configurations: ready-mades, assemblages, accumulations and environments. As stated, Kaprow sees how *junk* became a password for how things that had fallen by the wayside, whether old tyres, cardboard boxes, buckets, or brooms, were reintegrated and given new life. In Gnostic terms, the artist parodies or inhabits the role of demiurge reanimating matter: the object is played with, reinvented, recast, and reshaped. Allan Kaprow's 'Environments' from the late 1950s were constructed out of junk and everyday materials, exploding the distinctions between high and low art materials. Both Kaprow and Oldenburg turned to the streets, unconventional sites, and the accumulations of garbage on the streets of New York. Oldenburg's criticism of the material nuances of high art were reinvigorated by discarded things as revealed in his statement that 'a refuse lot in the city is worth all the art stores in the world.'[50]

Fallen-ness and fallen matter were the methods and metaphors in a broader enterprise of negation. The endless cycle of commodity and waste that was the by-product of burgeoning consumerism plays out through a Gnostic sensibility that stalks art. In privileging commodity over junk, matter is shaped to use-value and symbolic meaning in the commodity-sign; however, discarded objects and trash are shape shifters that haunt the commodity. From 1947 onwards, a new landfill site was opened on Staten Island absorbing vast amounts of commercial and residential waste before its closure in 2002. By 1986, 'Fresh Kills' (a subject for artist Mierle Laderman Ukeles) was the largest dump in the world, absorbing up to 13,000 tons of domestic waste per day. In 1972, Gordon Matta-Clark's

---

48   Joshua Shannon in Dezeuze, *Almost Nothing*, 53.
49   Schulz, *The Street of Crocodiles*, 59–60.
50   Claes Oldenburg in Anne Rorimer, *New Art in the 60s and 70s: Redefining Reality* (London: Thames and Hudson, 2001), 31.

short 16-mm film (*Fresh Kill*) recorded the artist destroying his dump truck with a bulldozer on the site. Towards the end of the 13-minute film, Matta-Clark cuts together multiple sequences showing the dismembered vehicle being tipped down a mountain of rubbish. We can now link a series of entropic moves together from Pollock's paint pouring to Smithson's giant pours such as *Asphalt Rundown* to Matta Clark's film of pulverized matter. Trash gave life to sculpture. Its abundant availability on both the streets of Manhattan and in the garbage hills around the suburbs opened sculpture to infinite extension and possibility. Easy metamorphoses between commodity-use and obsolescence inspired artists such as Claes Oldenburg and Allan Kaprow to explore the scrap material created by consumption. If agency is seen to lodge more vividly in impersonal matter, then this awareness is more keenly felt in a late industrial age with its rubbish heaps of redundant matter. It is accompanied by the notion that bodies themselves are part of the predicament of stranded matter in the modern age. In her book, *Vibrant Matter*, Jane Bennett reflects on how rampant consumerism has occluded our awareness of the vitality of matter. On examining a junk heap, she describes her reaction:

> It hit me then in a visceral way how American materialism, which requires buying ever-increasing numbers of products purchased in ever-shorter cycles, is *anti*materiality. The sheer volume of commodities and the hyperconsumptive necessity of junking them to make room for new ones, conceals the vitality of matter.[51]

Junk food that emerged in the early 1950s, together with its cheap, throwaway packaging was significant for the subject matter of Pop. Oldenburg's *Floor Hamburger* and *Two Cheeseburgers with Everything (Dual Burgers)* reflect his sculptural sympathies: 'I am for an art that takes its form from the lines of life itself, that twists and extends and accumulates and spits and drips, and is heavy and coarse and blunt and sweet and stupid as life itself.'[52] In their squat and sprawling condition, these floor sculptures possess

---

51   Bennett, *Vibrant Matter*, 5.
52   Rorimer, *New Art in the 60s and 70s*, 31.

a sort of recalcitrance and lassitude like some gestural effect of the body. In his analysis of Oldenburg's poetics of softness, Max Kozloff remarks:

> Juxtaposed solely with their environment, Oldenburg's works act as a harsh social indictment. They subvert every premise upon which the cajoling, commercial culture is built; they are monumental bad news, emphasizing the unworkability and the obsolescence of machines, and, above all, a lack of control over matter that American society is pledged not to mention.[53]

Behind the spectacle of hyper-consumption and obsolescence is its collapsed matter that speaks of those hidden forces that move across the social and commercial world without limit or reason. Within the context of Gnostic doubt, Oldenburg imbues matter with a 'creaturely squirming', as if nothing of contemporary materialism can be trusted to remain fixed in identity.[54] Gravity connects sculpture to the body: objects resemble or move like bodies, whilst bodies are turned into things. By enlisting soft materials, such as felt, latex, and *papier maché*, Morris and Oldenburg incorporate bodily sensations and effects of drooping and hanging to reveal the action of gravity. Morris spanned a transition from Minimalist rigidity to soft anti-form works, informed by his use of dance and the body. His interest in heavy industrial felt from 1967 on formed part of this sensory matrix between hard and soft, since the weight and drag of the material made visible through cutting and hanging felt belonged to the repertoire of bodily gesture.

Oldenburg turns rigid objects such as light switches and toilets into floppy obsolescence.[55] Objects are anthropomorphized through their softness and their pendulous states, reminiscent of our own shifting sensations of tension and relaxation. Sculpture in this period was awake to the softness of the body with its pliancy, demonstrating how weight deforms matter. Benjamin Buchloh describes the use of gravity in Oldenburg's work:

> Oldenburg … had taken the reduction of plastic phenomena to its natural origin: the system of coordinates formed by gravity and the temporal-spatial continuum

53 Kozloff, *Renderings*, 232.
54 Kozloff, *Renderings*, 233.
55 Max Kozloff, *Renderings*, 228.

where gradual processes involving masses and relative forces become plastic events, as seen in his 'Soft Objects.'[56]

Alternatively, soft, drooping states make dead things seem alive, just as in Schulz's short treatise, 'the whole of matter pulsates with infinite possibilities.'[57] In a throwaway culture, the junkyard or scrapheap were spaces of creative promise for artists, resisting a commodity cycle that was out of control.

Force Fields

Susan Leigh Foster argues that by expanding our concept of the word, choreography, we illuminate the hidden ways in which the body is shaped within different frames or networks of power. By exposing these invisible threads, artists and performers critique the body's normal activities and routines as forms of authority control. These tacit agreements between body and gravity subvert normative readings of space, movement, and behaviour. Judson performance took nothing for granted. Gravity, as democratic force field, is consciously reified in order to shake off the idea of the body as a creature of habit, thus establishing common ground between performer and audience. Artists, sculptors and performers started to present work in non-traditional venues, opening up new potential arenas for display in city streets, shop windows, overlooked urban lots, roof tops, walls and even ceilings. Speaking of her *Equipment Pieces* from the early 1970s, Trisha Brown writes: 'I always feel sorry for the parts of the stage that aren't being used. I have in the past felt sorry for the ceilings and walls. It's perfectly good space, why doesn't anyone use it?'[58] Using exterior walls and rooftops as spaces for

---

56 Benjamin Buchloh, 'Process Sculpture and Film in the Work of Richard Serra' in *Richard Serra*, ed. Hal Foster and Gordon Hughes (Cambridge, MA: The MIT Press, 2000), 3.
57 Schulz, *The Street of Crocodiles*, 59.
58 Brown in Banes, *Terpsichore in Sneakers*, 81.

performance, Brown set up a foil to how the body was normally corralled through pedestrian city space. Just as Matta-Clark subverted orthodox thinking about architecture, as examined in Chapter 1, so Brown challenged architecture by undoing the normal logic of how the body operates through the gravitational field. Her disregard for the conventions of gravitational space is recalled in Astaire's 360-degree dance in *Royal Wedding* and Martin Kersels's parody of Astaire in *Tumbleroom* (2001).

Unless we accidentally fall, our common sense of gravity's pull is generally subdued. Formulated as a law in the seventeenth century, gravity was incorporated into the realm of science and, thus, its mysterious action was accommodated into comforting habits of measured space and uprightness. By testing the assumptions of orthogonal space, Trisha Brown opened up radical new readings of space. *Walking on the Wall* of 1971, re-performed at the Barbican's exhibition, *Pioneers of the Downtown Scene* in 2011, subverts gravity-ordained readings of space by choreographing a performance on the walls, suspending the performers in harnesses. Tied to ropes, the walkers aim to move in normal pedestrian motion, moving around each other when necessary, their bodies pulled out at 90 degrees to the wall. In other performances created by Brown, performers walked down the side of buildings or moved on rooftops in New York's SoHo. This disregard for how architecture trains or moves the body is also evident in Gordon Matta-Clark's cut sections through buildings, walls, and floors, which erode these rigid frameworks of wall and ceiling, both physically and metaphorically.

The discourse of the puppet produces a kind of ironic anthropomorphism that reflects the position of humankind controlled by hidden forces. In so many performances and artworks of this period, gravity is used as a test bed. It is employed as a prime metaphor for a counter politics, standing in for those invisible networks that enclose the body. This might be analysed via a metaphor connected to Erving Goffman's concept of frame analysis.[59] Goffman's frame analysis constructs an image of the individual as puppet or actor operating within social, political or environmental frameworks, forced to act or behave in accordance with pre-defined

---

59   Erving Goffman, *Frame Analysis: An Essay on the Organisation of Experience* (Boston: North Eastern University Press, 1986).

expectations. This view of human beings sees them as heavily programmed or encoded, reducing any conviction that they might possess their own will and agency. These rhetorical frames filter perceptions of society so that the actor-individual is encouraged into particular interpretations and perceptions. If these frames subject the individual into codes and habits of thinking, then the puppet theatre might provide an apt metaphor for this idea of framing. Allan Kaprow cites Erving Goffman's sociological study *The Presentation of Self in Everyday Life* in his essay, 'Participation Performance', in which Goffman argues that everyday routines are highly coded performances of a kind.[60] As Kaprow states: 'Human beings participate in these scenarios, spontaneously or after elaborate preparations, like actors without stage or audience, watching and cuing one another.'[61] When choreographers such as Rainer, Paxton, and Brown reveal the body in its submission to the law of gravity, this is to rework the physical condition of its pull into the realms of the political, social, and ethical. Kaprow's interest in frames is primarily an interrogation of how art and life blur at the edges with the boundary between artist and audience eroded in terms of participation and performance.

The politics of participation has been an increasingly prominent feature of art, a critical method of resisting authority and regulation. In the 1960s and early 1970s, circus or playground props were integrated into art and performance, used by both audience and artist or performer. In 1960, Simone Forti first performed *See-Saw* at the Reuben Gallery in New York, a space co-founded by Allan Kaprow in which two performers, male and female, balance on a plank across a saw-horse, shifting their weight back and forth. Reminiscent of Wood and Harrison's games of double agency, her powerful play with gravity in *See-Saw* led on to *Slant Board* of 1961 (*Five Dance Constructions and Some Other Thing*) in which performers or members of the audience would mount a ramp, pitting their exertions against the force of gravity. In 1971, Robert Morris installed his choreographed exhibition, *Bodyspacemotionthings*, with the audience encouraged to move across a set of apparatuses. Parallel bars, slanted surfaces, tunnels, a tightrope and

60   Erving Goffman, *The Presentation of Self in Everyday Life* (London: Penguin, 1990).
61   Kaprow, *Essays on the Blurring of Art and Life*, 187.

see-saws were the props used to engage and involve the audience. Morris rejected the classical illusion of transcendence from the body, disguised in light *tulle* and wooden blocks: 'From the beginning I wanted to avoid the pulled-up, turned-out, anti-gravitational qualities that not only give a body definition and role as "dancer" but qualify and delimit the movement available to it. The challenge was to find an alternative movement.'[62] Morris's exhibition at Tate was closed when someone fell off the tightrope, but his reinvigoration of circus or playground movement signalled a key shift in the relationship between performer-audience and subject-object.

Such relations are evident in an early performance, *Column* (*c.* 1961). For this, Morris constructed a human-scaled column out of plywood and painted grey. His original intention was to stand inside the column and then make it collapse; however, the artist injured himself in rehearsal and so used a piece of thread to pull the empty column down in the performance. With the column standing in for the body, its collapse appeared spontaneous and automatic. First 'performed' at the Living Theater in New York, *Column* was upright for 3.5 minutes, then prone for another 3.5 minutes. Morris's falling column institutes a correlation between architectural form and our condition of uprightness. Morris had previously taken instruction with his wife, Simone Forti at Anna Halprin's workshops on the West Coast and in one class produced a sculptural quality to the way his body interacted with the ground. As Forti recalls:

> I remember vividly the movement Bob did on this particular day. He had observed a rock. Then he lay down on the ground. Over a period of about three minutes he became more and more compact until the edges of him were off the ground, and just the point under his centre of gravity remained on the ground.[63]

Morris provides a crucial link from Minimalist objects to soft sculptures, forged through his involvement in dance and mediated partly through his felt works from 1967 onwards. Moving felt behaved more like the body. By slashing and hanging the areas of felt to the wall, he allowed forms and

62   Robert Morris in Rosenthal, *Move*, 13.
63   Rosenthal, *Move*, 13.

movements to arrive indeterminately out of the material, rather than pre-conceiving the final form. This was for him, 'a direct revelation of matter itself.'[64] Morris was careful to acknowledge the silent partner in his hung felt works: 'Sometimes a direct manipulation of a given material without the use of any tool is made. In these cases considerations of gravity become as important as those of space. The focus on matter and gravity as means results in forms that were not projected in advance.'[65] Morris's *Untitled* (*Scatter Piece*, 1968–1969) was originally shown at Leo Castelli's and after Castelli's death was stored in a backroom. Subsequently taken to be rubbish, it was moved out and thrown on a scrapheap. Morris imagines the metal and felt now oxidizing and with the felt of his work lining the burrow of an animal. He describes his work's now chthonic existence as a ghostly or gothic double of the work reinstalled in 2010. The work occupies a space of indeterminacy in its original incarnation, 200 pieces of metal and felt scattered randomly in various spaces with half a dozen or so piled up in a corner. By reinstalling *Scatter Piece*, Morris feels the remade work to be a fraudulent double or imposter, but at the same time and with redoubled energy, this radicalizes the idea that order can be imposed on matter. Will and agency are made subservient to entropy and randomness. As in Fischli and Weiss's *Equilibrium* series, gravity automates the process of making at a fundamental level. For Morris, agency reduction is the hidden narrative mobilizing art in the 1960s and 1970s. Operating within the trail fumes of American industrialization, he questions our assumption of control over matter.

In terms of radical practices of the 1960s and 1970s, the blind and ineluctable force of gravity attracted artists and performers to stray beyond conventional boundaries, aesthetically and politically. In Kleist's motif of the puppet theatre, gravity, grace, and the fall are recognized as elements of Gnostic doubt that still resound in modern and contemporary life. Kleist's vision corresponds with the way performers such as Morris, Rainer and Oldenburg reanimate those commonalities of matter, muscle, and mind

64   Suzann Boettger, *Earthworks: Art and the Landscape of the Sixties* (Berkeley and Los Angeles: University of California Press, 2002), 97.
65   Boettger, *Earthworks*, 99.

into forms of resistance and into fantasy conversions between bodies and objects, trash and art. Gravity produces chance effects, displacing agency from the artist as authorial centre. Work or effort, as explored here, is centred on probing the natural physics of materials and muscles beyond the tired idiom of classical grace. No expression, no illusion, no spectacle, just the literal fact of mass and muscle or as Yvonne Rainer described her dance: 'a ten-ton truck stuck on a hill'.[66] Put another way, momentum gained momentum.

Gravity, as considered through these radical performances, dances and sculptures, reveals the ground as a quiet, democratic partner, a level playing field, which anchors temporary habitations and doomed transports into the air. In fact, its very democracy in action was matched metaphorically to more even-handed readings of the body. Democratic processes were important at the Judson Dance Theatre with the use of untrained dancers, freedom in experimentation, and an egalitarian approach to dance as concept. Mass became critical as political issues were screened through the shifting co-operations of force and body. Rainer characterized her dance movements as 'work-like' in that they operate as images of labour, but at the same time, she shifts the after-image of the factory into one of choice and freedom. Catherine Wood writes: 'In Rainer's conception of the mind as muscle, the acting-out of that "manual labour" and "everyday work" is provocatively presented as dance, a decadently non-productive activity.'[67] As Steinberg argues, the principle of work is often assimilated to the American ethic of hard work and efficiency. Variegated textures of motion and falling show how we cavort or suffer with the body, accepting or escaping the drag of our own weight. But as machines continue to surpass the human body's ability to lift and pull, where does this leave the discourse of the puppet? In the automaton-like movements of Rainer's dance, Oldenburg's floating figures of *The Street*, and Brown's suspended walkers lurks a vision of the human being as puppet. Kleist's essay about the puppet theatre and Schulz's Gnostic vision of matter enclose deep fears about our fallen state and loss of will, which are more relevant than ever in

66   Rainer in Lepecki, *Dance*, 48.
67   Wood, *Yvonne Rainer*, 87.

light of how technology's automatism now shadows life. John Gray reminds us of our contemporary position within Gnostic thought:

> Many people today hold to a Gnostic view of things without realizing the fact. Believing that human beings can be fully understood in the terms of scientific materialism, they reject any idea of free will. But they cannot give up hope of being masters of their destiny.[68]

Kleist teaches us to envy the condition of the puppet by learning its true condition of helplessness and dependence on a puppet master. Schulz's Gnostic reading of Kleist twists the knife deeper so that the lightweight trash and sawdust coiled into the heart of the puppet becomes a rich but terrifying vision of matter, impervious to human will and control. Shadowed by the phantasmagoria of commodity culture, garbage heaps, and dilapidated streets in and around 1970s New York, artists saw possibilities in the discarded and the obsolete. Although producing art and performance out of the accidental fall looks like freedom, it actually signals a retreat into doubt and into a dark vision of our ultimate dispossession by machines and urban automatism. If gravity infiltrates this story of agency reduction in art, it operates through the rhetorical figure of the fall, a real (though a weak) force that is shadowed by the power of product accumulation, mechanical reproduction, and the market.

68   Gray, *Soul of the Marionette*, 9.

Afterword

As I complete this book, Cornelia Parker has just been named official artist for the June 2017 election. On a visit to the House of Commons, Parker toured the building, examining and photographing the green leather seats with their creased imprints from the weighty behinds of various ministers and members, as if they were deflated whoopee cushions. She describes her temporary role within these seats of power, as being that of a court jester, who possesses the immunity to comment critically on politics. In an interview on BBC Radio 4's *Today* programme with Mishal Husain (1 May 2017), Parker talked about how she sees politics in everything, even the debris on her kitchen table. Her sensitivity to materials and the environment is ecological and alchemical, evident in the way she transforms objects and matter. Whether interrogating the link between cosmic and terrestrial forces in her work on meteorites, or dropping lead words from clifftops, Parker captures and freezes gravity through different processes and materials. Gravity has been consistently and consciously evident in her sculptural practice. As she explains: 'My work is consistently unstable, in flux; leant against a wall, hovering, or so fragile it might collapse.'[1] As Parker starts to travel across the country during the election campaigns, I draw on her sculptural practice as a useful guide with to summarize the main ideas in this account of gravity's role in contemporary art.

Parker's sculptural imagination connects detailed, speculative approaches to material combined with concepts of mass and gravity. How does her sustained interest in gravity reflect the key themes and ideas of this book? Gravity is made vivid in her continual iterations of ground and earth, through activities of excavation and levitation; in her alchemical readings of mass and metal; in her co-operation with the heavy work of industry, and in popular cultural references, such as the 'cartoon deaths' of

---

1   Iwona Blazwick, *Cornelia Parker* (London: Thames and Hudson, 2014), 107.

her objects or references to Hollywood film; gravity is also evident in her critical use of the word, 'mass' in many of her sculptural installations. She uses dynamite, steamrollers, and hydraulics to change objects and materials into mangled tautologies and caricatures. Her experiments with the mass of lead, silver, and brass mobilize a poetics of metal, the resistance of the material she works with providing the provocation for violent transmutation. Working with metal means that her creative tools often belong to the world of manual labour and heavy stuff: gun factories, an industrial 250-ton press, and a steamroller. Metal's relationship to mining not only concerns the history of heavy industry, but also underwrites the contract with dematerialized technologies, such as the coltan used in mobile phones. Just as gravity dilates time, so making a dent in mass demands patience and *longueur*. Objects emerge as if swollen out of the force produced by space-time.

Parker combines creative approaches to material with more visionary concepts of space and mass. In *Words that Defy Gravity* of 1992, she cast a lexicon related to the word gravity in lead and threw the cast words off the Dover Cliffs, held back from the edge by a belt tied around her waist pulled on by two friends. A certain theatricality pertained to the process of making this work with the words tossed exuberantly over the edge, just as Yves Klein had famously thrown himself out of an upstairs window. This moment was photographically frozen with the word 'specific' hovering in the air just before falling over the drop, just as Klein was captured in mid-flight.[2] These crushed words, transformed by their fall from height, were then suspended in a gallery just above the floor. The words are tautological, related to the meaning of gravity, but ultimately destroyed by it. Parker admits to being afraid of heights, yet is simultaneously attracted to the abyss. In *Measuring Niagara with a Teaspoon* (1997), the height of Niagara Falls is extruded into a thin wire made from a single Georgian silver spoon, its sublime grandeur contained in a tiny domestic object, just as Newton bridged the scale between the celestial and terrestrial mechanics.

2   Blazwick, *Cornelia Parker*, 55.

There is a redemptive logic at work in Parker's creative ideas about gravity, whereby destroyed objects are given a poetic afterlife, sometimes literally salvaged from the ashes. In *Anti Mass* (2005), Parker takes the charred remains of a Southern Baptist church of a largely African American congregation, after a sinister arson attack, and suspends the fragments so that they float in air like her famous exploded garden shed (*Cold Dark Matter: An Exploded View*, 1991). The word 'mass' has variant meanings. Its religious meaning exists in reference to the sacrament held aloft, but is also invoked in its meaning as dense and weighty matter. Mass is also the drawing together of a congregation of people within a sacred space. Parker's sculpture is alchemical; it is as if she quietly inhabits the role of demiurge, penetrating the very depths of matter and shifting its densities. By doing so, she exhibits a respect for matter that is naturally coextensive with an ecological sensitivity. Her ideas enlarge and particularize terrestrial matter, imbuing it with the same cosmic wonder as meteorites or dark matter.

Parker's *Cold Dark Matter* connects the cosmic with the humdrum and humble quality of the garden shed. Parker organized the detonation of a garden shed and all its contents, working with the British Army School of Ammunition at Banbury. The shed was a composite of other sheds and its contents were gleaned from car boot sales or donated by friends from their own sheds. By exploding the shed, Parker references the violence of the comic strip or action film, as well as alluding to the Big Bang. The work, combining sculpture, film, performance, and participation by other agencies, was included in Tate Modern's inaugural exhibition, 'Between Cinema and a Hard Place' in 2000, after its original commission by the Chisenhale Gallery in London. By suspending blown-up objects and shed fragments, Parker recreates the shed as mini universe with its 'sun' light bulb. She notes that when suspended, the destroyed objects started to 'lose their aura of death and become reanimated.'[3] In the subtitle of *Cold Dark Matter*, 'An Exploded View', Parker recalls engineering manuals and exploded view illustrations that take objects apart as if they are blown out from their gravitational centre to reveal their inner workings. Manuals devoted to mechanics

3   Blazwick, *Cornelia Parker*, 50.

and engineering are directly descended from Newtonian physics. Utilitarian followers of Newton, such as Jean Desaguliers promulgated terrestrial physics through guides for the artisan and engineer, replacing tacit knowledge and rules of thumb. From the seventeenth century onwards, measuring and standardizing work helped to ground it in a 'rational' relationship between physics and exertions of the body, as explored in Chapter 3. Parker teases out the creative possibilities of technical know-how, extending agency to organizations representing law and order, or those with practical skills and machines, such as gun manufacturers or steamroller operatives. By co-operating with the army in *Cold Dark Matter*, Parker quietly subverts the military's association with law and order.

As in the 'cartoon death', the remains of the shed are brought back to life with renewed agency. There is a strange combination of diabolical violence, often found in comics and cartoons, with an ethical concern for how we treat things that share our terrestrial condition. Sometimes, her objects meet a 'cartoon death' in a squashed condition of the pratfall like that of the old slapstick comedy. However, the frenetic tempo of the cartoon is held in suspended animation. Parker dilates the time of an explosion into long duration. In Bruno Schulz's short treatise on 'Tailors' Dummies', 'lifelessness is only a disguise behind which hide unknown forms of life'. This statement is instrumentally vivid throughout Parker's art as matter shifts shape into new and varied forms of existence. Schulz's treatise extols the Gnostic spirit, characterized by giving 'priority to trash', or being entranced by junk's downtrodden or shabby appearance as explored through the junk aesthetic. By lighting the sculpture from the centre, Parker casts shadows onto the wall, projecting 'a shadow puppet show in which the objects hover with subliminal intent.'[4] By extending the split second of an explosion, she freeze-frames gravity with objects illuminated in a state of suspended animation. The light bulb at the centre of *Cold Dark Matter* provides a laconic statement on how the celestial and terrestrial worlds are locked together in blind symmetry. This connection between gravity's mysterious source, causality, and the Gnostic vision is quietly evident throughout many

4   Blazwick, *Cornelia Parker*, 14.

of the practices I have referred to; evident in Robert Smithson's description of crystalline accretion, Technicolor Hollywood or 'smoking mirrors' that signify dark demiurgy; present also in the way bodies and objects in Stan Douglas's photographs (*Midcentury Studio*, 2010) exist in a *mise en abyme* of gross matter with floating oranges, knives, and rings levitated in the camera's flashbulb.

We might say the garden shed is a heterotopia of sorts, a space of alternative inhabitation and occult activity where objects are brought together in strange encounters. A demiurge instinct moves artists to play in mock laboratories, suburban garages, and garden sheds as if the role of amateur scientist or engineer were more fitting to their activities. Within these spaces for play and experiment, things collide, fall, and fly through the membrane of gravity. Gravity is invisible mover, hidden cause, and the obtuse reality of matter. It is therefore embedded in the very heart of creative play, evident in the figure of the demiurge in both Platonic and Gnostic iterations. The Platonic version of the demiurge describes a benign artisan-like creator, who fashions the world in ideal ways, whereas in its Gnostic aspect, the demiurge is a diabolical figure, who has moulded a world of dull and heavy matter to imprison the human spirit. So many of the contemporary artists discussed here seem unconsciously guided by an awareness of how contemporary life is tempered by a Gnostic spirit of doubt and paranoia; however in a reversed reading, they seem to reclaim the power and poetics of being gravity-bound. I think here of Bas Jan Ader's small experiments in falling or the beautiful demonstrations of entropy and falling in *The Way Things Go* by Fischli and Weiss. With their creative energy aimed at altering our habits of perception, these artists reconfigure our relationship to the material world, showing how to lessen our control of it. They uncover the potential beauty of the fall. By tipping, balancing or dropping objects, they revalue redundant waste and everyday objects by tinkering, reassembling, and transforming things to recover the latent energy and liveliness they harbour. In a form of reverse sympathetic magic, artists show how we might retard our orchestration of the object world, to recover the agency of matter.

Parker's work is symptomatic of a post-human philosophical tendency in the way obsolescent things are revalued. As examined in my final chapter, artists such as Allan Kaprow, Claes Oldenburg, and Gordon Matta-Clark were transfixed by gross accumulations of waste and discarded objects in America's commodity boom of the 1950s, just as the industrial ruins and defunct mines around New Jersey engrossed Robert Smithson. The staggering mountains of waste that built up on the scrapheaps and landfills around New York seemed to indicate a radical and painful alienation from the world. Parker's appropriation of rocks and earth in works such as *Neither From Nor Towards* (1992) or *Edge of England* (1999), though wholly different in orientation to the earth art of Robert Smithson, still magnifies the issues of post-human ecology. She cites earth artist Walter De Maria as an influence through identifying the specific origins of earth, such as the substrate extracted from under the Leaning Tower of Pisa in *Subconscious of a Monument* (2001–2005). This was the site of Galileo's experiments with gravity and is now an urgent subject for engineers attempting to reverse its collapse. Parker was able to obtain the earth excavated from underneath the tower, in order to subside the monument in the opposite direction to correct its lean. Suspending the fragments in an installation, she defies gravity twice over. In *Neither From Nor Towards* (1994), she created a sculptural installation composed of old bricks from a row of houses that had fallen over the edge of a cliff near Dover. Over time, the bricks had been washed by the waves to resemble pebbles. The fragments were then suspended on wire in the formation of a house, creating the illusion of the house hovering mid-fall or else resurrected after its collapse. As explored in my first chapter, architecture is subtended by the promise made by Newtonian gravity, of certainty, predictability, and stability, these in turn being converted into habits of assumption. In my selections, houses fall off cliff edges, woodsheds collapse and suburban homes are tilted back or float off like the tumbling house in *The Wizard of Oz* to unseat the certainties of architecture and reveal the naked contingency of being gravity-bound. Buster Keaton's films were the perfect encapsulation of how architecture is merely a flimsy pretence of shelter and uprightness. Keaton's simple clapboard houses spin and fall around him, leaving his body exposed to the whims of nature. Gordon Matta-Clark was to recall the antic freedom of early film comedy after splitting a whole house in two (*Splitting*, 1974). Keaton's deadpan expression and mute passivity, echoed in Steve McQueen's

Deadpan (1997), is an intuition of how the body, like the earth, is power-less to resist its continuous declination.

In a more recent work, Parker alludes to Hitchcock's 1960 horror film, *Psycho. Transitional Object (PsychoBarn)*, a Roofgarden commission by the Met installed in 2016, which combines references to the Bates's gothic man-sion, the classic American red barn and paintings by Edward Hopper. Nearly 30 feet in height, it is composed out of a deconstructed red barn, familiar as an image of wholesomeness, which is then uncannily subverted through its likeness to the house in *Psycho*. As in Hitchcock's set, it consists of just a façade and sidewall with scaffolding supporting it from behind, thus playing on ideas of presence and illusion. Its temporary occupation of the rooftop is also paralleled in the rickety, collapsible film sets of early Hollywood comedy and the high-rise burlesque comedies, such as those of Harold Lloyd. By subverting the modern Manhattan skyline with a building from America's rural and popular vernacular tradition, Parker illuminates the psychological co-ordinates of architecture and the space of the fall.

The low and the down of popular culture is an important conduit into aspects of contemporary art practice, as if the clichés and pathos of folk traditions might be preserved in certain ecologies of material and object. Falling, a consistent theme in this book, is our fundamental condition, recollected in the clown's pratfall, the suicide's leap, and death-defying tightrope walk. In *Thirty Pieces of Silver* (1988–1989), Parker refers to the violence done to bodies in cartoons. Taking thirty silver objects, often con-nected to commemoration in people's lives, such as silver plates and coins, she flattened the objects with a steamroller and then suspended them in circular formations above the ground. As she claims: 'I took inspiration from my childhood love of the cartoon "deaths" of Roadrunner or Tom and Jerry, who were constantly being shot full of holes, blown up by dynamite, run off cliffs or flattened with a shovel.'[5] The title refers to the betrayal of Christ by Judas that led to his death and resurrection. Silver has resonances of lightness and gloss, the silver lining of hopes and dreams, here deflated and then resurrected. The material is perhaps also recovered in the term,

5   Parker in Blazwick, *Cornelia Parker*, 37.

'silver screen' to describe the association of Hollywood with glitz, derived originally from its use in the lenticular screen of early film projection. Parker uses silver both for its reflective and opaque qualities. Mirrored surfaces, in Smithson's view, are darkly Gnostic: the celestial ceiling shines back at us, the crystalline Deco interior is an airless prison and the gaudy, artificial colours of Technicolor film are emanations or illusions issuing from the demiurge artisan.

Switching between poles of the animate and inanimate, Parker's figure of the fall is part of what she describes in her work as 'embedded choreography'.[6] As Bruce Ferguson says of her sculpture, 'her objects and installations do seem to exist in a particular moment of a process, like newly costumed performers in a dance, actors with or without a speaking part, or ventriloquists' dummies playing a duplicitous role.'[7] States of falling and floating produce a theatrically charged diagram of force. Her sculpture preserves a stubborn resistance to dematerialization, magically reawakening the potential for objects to float or collapse, often against a stopped clock. Her suspended moments like the split-second explosion in her detonated shed suggest a momentary freedom from gravitational force and simultaneously, entrapment. Her work enlivens those incompatible demands in performance between metaphysical aspiration and physical resistance. Her spectacular levitations recall Adorno's description of the circus trick as based on the dialectic of impossible ideal and senseless exertion. What Adorno additionally found in circus's empty virtuosity was an image of aesthetic redemption, cleansed of subjective agency, stable social identity and authoritarian control. Parker's objects dangling and dancing on wire are literally and metaphysically salvaged. Even the tedium associated with some of her material processes can be measured against the rote mechanics and repetitions of circus illuminated by Adorno as the very enigma of art – in both cases, why waste all that effort?

In this account of how artists have more consciously negotiated the force of gravity, I have followed its myriad effects within art and

6   Parker in Blazwick, *Cornelia Parker*, 14.
7   Ferguson in Blazwick, *Cornelia Parker*, 14.

performance: the actions and results of falling paint, falling bodies, lean-ing bodies, flying objects, spinning machines, heavy lifting, and demateri-alizing media. Through my examples, contemporary artists have sequestered a terrain on which to unloose the mass of matter from its instrumental imperatives in order to play with it empirically, shining a tiny light in a big darkness. In fact, many of the works discussed in this book resemble early science experiments with objects or bodies made to spin or fall. Their experimental quality is fundamental to contemporary notions of creative practice, characterized as a desire for the open-ended, the provisional or the temporary, rather than by notions of success or finality. Lurking within the various guises and examples of the fall is a strain of anti idealism. By inhabiting an impulse towards anti grace, performers and sculptors have privileged trips and pratfalls that signal the potential beauty of doubt and fragility. They have assumed a type of artisanal modesty to mock the hubris associated with the 'demiurge spirit'. My account of how gravity has infil-trated artistic practices since the 1960s, as a sort of immaterial readymade, has exposed all manner of small lyrical encounters or temporary solutions between bodies and objects.

By speculating on the explicit or the sometimes half-conscious applica-tions of gravity in art, performance and sculpture, I have traced themes of work, play, and popular entertainment across diverse branches of the arts, such as dance, architecture, film, sculpture, and performance. By using an invisible force of wide effect such as gravity, artists and performers have necessarily focused on small and intense investigations into the substance and matter of everyday things, but also dispersed their activities across many domains. Harnessing gravity in their work, artists and performers have dis-covered new ways of dismantling the traditional boundaries of medium. Thinking gravity as a quasi-medium has allowed them to stretch their canvas to encompass the outer limits of earth's horizon, of earth, ground, sky and space. This enlargement of vision has in turn thrown the excesses of automatism and machine efficiency or the nightmare accumulations of commodity culture into sharp relief. The artist's appeal to his or her own vanishing mediation and control is also, with certain exceptions, config-ured in the broader agencies and coalitions used within artistic process or in the crossovers between studio and factory. Artists have magnified our consciousness of the matter and ontology of being earthbound, using gravity as medium, method, and message to show us how the local, global, and cosmic are intimately connected.

Gravity is a strange coefficient in time, retarding and producing a stay against its passage like a drop of paint, languorous but heavy with intent and potential effect. If, as Michio Kaku claims, gravity does not really exist, but is simply an effect of rippling space-time, then surely its role is one of vanishing mediation in our attempt to understand causality. By harnessing the invisible force of gravity, these contemporary artists and performers have exploited it as a vanishing mediator, downgrading their own subjective control in the act of creation, and by doing so, they have pursued a sensitive and ethical response to how we human beings are woven into the web of force and matter. This is part of the larger plot of agency reduction in which auteur subjectivity is considered a shade too hubristic, whilst in antithesis, falling awakens an ethical understanding of just how precarious life is.

Bibliography

Adorno, Theodor, *Aesthetic Theory*, tr. Robert Hullot-Kentor (London: Continuum Impacts, 2004).

Adorno, Theodor, and Horkheimer, Max, *Dialectic of Enlightenment*, tr. John Cumming (London and New York: Verso, 2016).

Alain-Bois, Yve, and Krauss, Rosalind, *Formless: A User's Guide* (New York: Zone Books, 1999).

Arnheim, Rudolf, *Film as Art* (Berkeley and Los Angeles: University of California Press, 1957).

Auiler, Dan, *Vertigo: The Making of a Hitchcock Classic* (London: St Martin's Press, 1998).

Bachelard, Gaston, *Earth and Reveries of Will: An Essay on the Imagination of Matter*, tr. Kenneth Haltman (Dallas: The Dallas Institute Publications, 2002).

Banes, Sally, *Terpsichore in Sneakers: Post-Modern Dance* (Middletown: Wesleyan University Press, 1987).

Barthes, Roland, *The Eiffel Tower and Other Mythologies*, tr. Richard Howard (Berkeley and Los Angeles: University of California Press, 1997).

Bataille, Georges, *Visions of Excess: Selected Writings 1927–1939*, ed. Allan Stoekl, tr. Stoekl with Carl L. Lovitt and Donald M. Leslie Jr (Minneapolis: University of Minnesota Press, 2004).

Batchelor, David, *Chromophobia* (London: Reaktion, 2000).

Batchelor, David, and Esche, Charles, *John Wood and Paul Harrison* (London: Ellipsis, 2000).

Battista, Kathy, *Renegotiating the Body: Feminist Art in 1970s London* (London: I. B. Tauris, 2013).

Bauman, Zygmunt, *Liquid Modernity: Living in an Age of Uncertainty* (Cambridge: Polity Press, 2000).

Beckett, Samuel, *The Complete Dramatic Works* (London: Faber and Faber Ltd, 1986).

——, *Murphy* (London: Faber and Faber Ltd, 2009).

Bell, Kirsty, 'Gordon Matta-Clark', *Frieze*, 78 (2003) <http://www.frieze.com/issue/review/gordon_matta_clark/>, accessed 8 March 2016.

Benjamin, Walter, *The Work of Art in the Age of Mechanical Reproduction*, tr. Harry Zohn <http://www.marxists.org/reference/subject/philosophy/works/ge/benjamin.htm>, accessed 2 June 2017.

Bennett, Jane, *Vibrant Matter: A Political Ecology of Things* (Durham: Duke University Press, 2010).

Bergson, Henri, *Laughter: An Essay on the Meaning of the Comic*, tr. Cloudesley Brereton and Fred Rothwell (Book Jungle, 2008).

Bickers, Patricia, 'Let's Get Physical' interview with Steve McQueen, *Art Monthly*, 202 (December 1996–January 1997) <http://www.artmonthly.co.uk/magazine/site/article/steve-mcqueen-interviewed-by-patricia-bickers-dec-jan-96-97>, accessed 2 February 2016.

Birnbaum, Daniel, and Volz, Jochen, eds, *Making Worlds* (*Fare Mondi*, 53rd Venice Biennale Catalogue), 2009.

Bishop, Claire, *Artificial Hells: Participatory Art and the Politics of Spectatorship* (London: Verso, 2012).

Boettger, Suzaan, *Earthworks: Art and the Landscape of the Sixties* (Berkeley and Los Angeles: University of California Press, 2002).

Brown, Helen, 'Simon Faithfull's "Gravity Sucks": Furniture's Giant Leap', *The Telegraph* (10 July 2009) <http://www.telegraph.co.uk/culture/art/art-features/5777571/Simon-Faithfulls-Gravity-Sucks-furnitures-giant-leap.html>, accessed 10 August 2012.

'Bruce Nauman, Clown Torture', Art Institute Chicago <http://www.artic.edu/aic/collections/artwork/146989>, accessed 19 August 2013.

Buchloh, Benjamin, 'Process Sculpture and Film in the Work of Richard Serra' in *Richard Serra*, ed. Hal Foster and Gordon Hughes (Cambridge, MA: The MIT Press, 2000).

Caillois, Roger, *Man, Play and Games*, tr. Meyer Barash (Urbana and Chicago: University of Illinois, 2001).

——, 'Mimicry and Legendary Psychasthenia' in *October*, 31 (Winter 1984), 16–32.

Calder, John, *The Theology of Samuel Beckett* (Richmond: Alma Classic Ltd, 2012).

Calvino, Italo, *Six Memos for the Next Millennium*, tr. Patrick Creagh (London: Vintage, 1996).

Canetti, Elias, *Crowds and Power*, tr. Carol Stewart (London: Phoenix, 2000).

Chown, Marcus, *The Ascent of Gravity* (London: Weidenfeld and Nicholson, 2017).

Cooke, Lynne, and Kelly, Karen, eds, *Robert Smithson, Spiral Jetty: True Fictions, False Realities* (Berkeley and Los Angeles: University of California, 2005).

Connor, Steven, 'Inclining to the View' (2010) <http://stevenconnor.com/inclining.html>, accessed 4 January 2017.

——, 'Man is a Rope', *Catherine Yass High Wire* (Artangel, 2008).

——, 'Next-to-Nothing', *Tate Etc.*, 12 (2008) <http://www.tate.org.uk/context-comment/articles/next-nothing>, accessed 18 February 2016.

——, 'Shifting Ground' (2004) <http://stevenconnor.com/beckettnauman.html>, accessed 23 January 2017.

Conrad, Peter, *Modern Times, Modern Places: Life and Art in the 20th Century* (London: Thames and Hudson, 1999).

Coole, Diana, and Frost, Samantha, eds, *New Materialisms: Ontology, Agency and Politics* (Durham: Duke University Press, 2010).

Cousins, Mark, *The Story of Film* (London: Pavilion Books, 2011).

Crary, Jonathan, *Suspensions of Perception: Attention, Spectacle, and Modern Culture* (Cambridge, MA: The MIT Press, 2000).

Cunningham, Merce, *Merce Cunningham Dancing in Space and Time*, ed. Richard Kostelanetz (New York: Da Capo Press, Inc., 1998).

Curtis, David, *A Century of Artists' Film in Britain* (19 May 2003–18 April 2004) <http://www.tate.org.uk/whats-on/tate-britain/exhibition/century-artists-film-britain/century-artists-film-britain-progra-22>, accessed 9 February 2016.

Davis, Janet M., *The Circus Age: Culture and Society under the American Big Top* (Chapel Hill: The University of North Carolina Press, 2002).

Dean, Tacita, 'W. G. Sebald', *October*, 106 (2003).

De May, Phillip, *Lucretius: Poet and Epicurean* (Cambridge: Cambridge University Press, 2009).

De Montaigne, Michel, 'When I play with my cat, who knows whether I do not make her more sport than she makes me?', *Apology for Raymond Sebond*, 1592 <http://www.sophia-project.org/>, accessed 1 May 2017.

Deleuze, Gilles, and Guattari, Félix, *A Thousand Plateaus: Capitalism & Schizophrenia*, tr. Brian Massumi (London: The Athlone Press Ltd, 1999).

Demos, T. J., 'Moving Images of Globalization', *Grey Room*, 37 (2009).

Dessauce, Marc, ed., *The Inflatable Moment: Pneumatics and Protest in '68* (New York: Princeton Architectural Press, 1999).

Dezeuze, Anna, *Almost Nothing: Observations on precarious practices in contemporary art* (Manchester: Manchester University Press, 2017).

Diserens, Corinne, ed., *Gordon Matta-Clark* (London: Phaidon Press Ltd, 2003).

Dumbadze, Alexander, *Bas Jan Ader: Death is Elsewhere* (Chicago: University of Chicago Press, 2013).

Enwezor, Okwui, 'From Screen to Space: Projection and Reanimation in the Early Work of Steve McQueen' in *Steve McQueen: Works 1993–2012* (Heidelberg: Kehrer Verlag, 2013).

Faithfull, Simon, <http://www.simonfaithfull.org/works/escape-vehicle-no-5/>, accessed 10 August 2012.

——, and Roberts, Ben, *The World Turned Upside Down* (Warwick Arts Centre exhibition: 4 October–14 December 2013) <https://www.warwickartscentre.co.uk/whats-on/2013/the-world-turned-upside-down-buster-keaton-sculpture-and-the-absurd/>, accessed 5 August 2015.

Fitzpatrick, Andrea D., 'The Movement of Vulnerability: Images of Falling and September 11', *Art Journal*, 66 (4), 84–102.

Flam, Jack, ed., *Robert Smithson: The Collected Writings* (Berkeley and Los Angeles: University of California Press, 1996).

Flusser, Vilém, *Towards a Philosophy of Photography* (London: Reaktion, 2007).

Foster, Hal, 'The Un/making of Sculpture' in *Richard Serra* (Cambridge, MA: The MIT Press, 2000).

——et al., eds, *Art Since 1900: Modernism, Antimodernism, Postmodernism* (London: Thames and Hudson, 2004).

Gates, Theaster, 'Zeitgeisters' [podcast] (21 July 2014), BBC Radio 4 <http://www.bbc.co.uk/programmes/b041469j>, accessed 8 August 2016.

Gillain, Anne, *François Truffaut: The Lost Secret*, tr. Alastair Fox (Bloomington: Indiana University Press, 2013).

Gilloch, Graeme, *Myth & Metropolis: Walter Benjamin and the City* (Cambridge: Polity Press, 1997).

Goffman, Erving, *Frame Analysis: An Essay on the Organisation of Experience* (Boston: North Eastern University Press, 1986).

Goffman, Erving, *The Presentation of Self in Everyday Life* (London: Penguin, 1990).

Graham, Rodney, 'Artist Statement, 2005', *Rodney Graham: Through the Forest*, ed. Friedrich Meschede (Ostfildern: Hatje Cantz Verlag, 2010).

——, 'Freud's Clinamen' in *Rodney Graham : Vancouver Art Gallery* (Vancouver: Vancouver Art Gallery, 1988).

——, 'Siting Vexation Island', *Island Thought: an archipelagic journal published at irregular intervals* (Brussels: Yves Gevaert Verlag, 1997).

——, 'Smithson's Brain' in *Smithson in Vancouver* (Vancouver: Vancouver Art Gallery, 2004).

Gray, John, *The Soul of the Marionette: A Short Enquiry into Human Freedom* (London: Penguin, 2016).

Gunning, Tom, 'The Cinema of Attraction[s]: Early Film, its Spectators and the Avant-Garde' in *The Cinema of Attractions Reloaded*, ed. Wanda Strauven (Amsterdam: Amsterdam University Press, 2006).

Güse, Ernst-Gerhard, ed., *Richard Serra* (New York: Rizzoli International Publications, 1988).

Harris, Steven, 'Pataphysical Graham: A Consideration of the Pataphysical Dimension of the Artistic Practice of Rodney Graham', *Tate Papers No. 6* (2006) <http://www.tate.org.uk/research/publications/tate-papers/06/pataphysical-graham-consideration-of-pataphysical-dimension-of-artistic-practice-of-rodney-graham>, accessed 4 April 2016.

Hatherley, Owen, *Militant Modernism* (Hants: O Books, 2008).

Heiser, Jörg, 'The Odd Couple', *Frieze*, 102 (2006) <https://frieze.com/article/odd-couple>, accessed 24 January 2017.

Hell, Julia, and Schönle, Andreas, eds, *Ruins of Modernity* (Durham: Duke University Press, 2010).

Hite, Christian, 'The Art of Suicide: Notes on Foucault and Warhol', *October*, 153 (Cambridge, MA: October Magazine and MIT, 2015), 65–95.

Hoberman, J., 'The Lost Futures of Chris Marker', *New York Review of Books* (23 August 2012) <http://www.nybooks.com/daily/2012/08/23/lost-futures-chris-marker/>, accessed 7 February 2017.

Israel, Nico, *Spirals: The Whirled Image in Twentieth-Century Literature and Art* (New York: Columbia University Press, 2015).

Iverson, Margaret, *Beyond Pleasure: Freud, Lacan, Barthes* (University Park: Pennsylvania State University Press, 2007).

James, William, *Collected Essays and Reviews* (London: Longmans, Green & Co.,1920).

Kabakov, Ilya, and Emilia, 'The Palace of Projects' (24 March–10 May 1998) <http://www.artangel.org.uk/project/the-palace-of-projects/>, accessed 14 August 2012.

Kaprow, Allan, *Essays on the Blurring of Art and Life*, ed. Jeff Kelley (Berkeley and Los Angeles: University of California Press, 1993).

Kimmelman, Michael, 'Cross Sections of Yesterday: Gordon Matta-Clark Retrospective', *New York Times* (23 February 2007) <http://www.nytimes.com/2007/02/23/arts/design/23matt.html>, accessed 2 February 2017.

Kleist, Heinrich von, 'On the Puppet Theatre' in *Selected Writings*, ed. and tr. David Constantine (Indianapolis: Hackett Publishing Company Inc., 2004), 411–16.

Kozloff, Max, *Renderings: Critical Essays on a Century of Modern Art* (London: Studio Vista, 1970).

KunstsammlungNWR, 'Saraceno, Tomás-in orbit', [video] (uploaded 21 June 2013) <http://www.youtube.com/watch?v=ROqL-8h_7DM>, accessed 28 February 2016.

Lakoff, George, and Johnson, Mark, *Metaphors We Live By* (Chicago: University of Chicago Press, 1980).

Lingwood, James, *Catherine Yass High Wire* (London: Artangel, 2008).

McEvilley, Thomas, 'Yves Klein and Rosicrucianism' in *Yves Klein 1928–1962: A Retrospective* (Houston, TX: Rice University, Institute for the Arts, 1982).

McKee, Francis, 'I saw a city in the clouds' in *Catherine Yass High Wire* (Artangel, 2008).

McLuhan, Marshall, *The Mechanical Bride: Folklore of Industrial Man* (New York: Vanguard Press, 2002).

Maddocks, Fiona, 'Double Vision' (19 May 2016) <www.royalacademy.org.uk/article/video-jake-chapman-studio>, accessed 1 April 2017.

Madsen, Jane, *Kleist and the Space of Collapse*, PhD thesis, UCL Bartlett School of Architecture, London, 2016.

Mehlman, Jeffrey, *Emigré New York: French Intellectuals in Wartime Manhattan 1940–1944* (Baltimore, MD: The John Hopkins University Press, 2000).

Meisel, Martin, *Realizations: Narrative, Pictorial, and Theatrical Arts in Nineteenth-Century England* (Princeton, NJ: Princeton University Press, 1983).

Meller, James, ed., *The Buckminster Fuller Reader* (London: Penguin, 1972).

Meschede, Friedrich, ed., *Rodney Graham: Through the Forest* (Ostfildern: Hatje Cantz Verlag, 2010).

Michelson, Annette, 'Bodies in Space: Film as Carnal Knowledge', *Artforum*, 7 (6) (February 1969), 54–63.

Millar, Jeremy, *Fischli and Weiss: The Way Things Go* (London: Afterall Books, 2007).

Miller, James, *The Passion of Michel Foucault* (Cambridge, MA: Harvard University Press, 2000).

Morris, William, 'A Factory As It Might Be', *Justice*, April–May 1884 <http:// www. marxists.org/archive/morris/works/1884/justice/12fact2.htm>, accessed 5 April 2017.

Morrissey, Simon, *Richard Wilson* (London: Tate Publishing, 2005).

Morton, Tom, 'Pale Kingdoms' in *Salla Tykkä: White Depths* (Bologna: Damiani, 2011).

Müller-Sievers, Helmut, *The Cylinder: Kinematics of the Nineteenth Century* (Berkeley and Los Angeles: University of California Press, 2012).

Mullins, Charlotte, *Rachel Whiteread* (London: Tate Publishing, 2004).

Nietzsche, Friedrich, *Thus Spoke Zarathustra: A Book for Everyone and No One*, tr. R. J. Hollingdale (Harmondsworth: Penguin, 2003).

Noever, Peter, ed., *Chris Burden: Beyond the Limits* (Ostfildern: Cantz Verlag, 1996).

Nye, David, *American Technological Sublime* (Cambridge, MA: The MIT Press, 1994).

Osborne, Peter, *Anywhere or Not at All: Philosophy of Contemporary Art* (London: Verso, 2013).

Paxton, Steve, 'Fall After Newton' in *Dance: Documents of Contemporary Art*, ed. André Lepecki (London: Whitechapel Gallery; Cambridge, MA: The MIT Press, 2012).

Quinodoz, Danielle, *Emotional Vertigo: Between Anxiety and Pleasure*, tr. Arnold Pomerans (London: Routledge, 1997).

Rancière, Jacques, *Aisthesis: Scenes from the Aesthetic Regime of Art*, tr. Zakir Paul (London: Verso, 2013).

Reynolds, Ann, 'Enantiomorphic Models' in *Robert Smithson*, ed. Jane Hyun (Los Angeles: Museum of Contemporary Art and University of California, 2004), 137–40.

'Richard Wilson's Hang on a Minute Lads I've Got a Great Idea Goes to Hong Kong', De La Warr Pavilion <https://www.dlwp.com/richard-wilsons-hang-on-a-minute-lads-ive-got-a-great-idea-goes-to-hong-kong/>, accessed, 28 February 2017.

Ridout, Nicholas, *Stage Fright, Animals and Other Theatrical Problems* (Cambridge: Cambridge University Press, 2006).

Rimell, Victoria, *The Closure of Space in Roman Poetics: Empire's Inward Turn* (Cambridge: Cambridge University Press, 2015).

Rorimer, Anne, *New Art in the 60s and 70s: Redefining Reality* (London: Thames and Hudson, 2001).

Rosenthal, Stephanie, ed., *Move: Choreographing You* (London: Hayward Publishing, 2011).

Sanders, James, *Celluloid Skyline: New York and the Movies* (New York: Alfred A. Knopf, 2003).

Schwartz, Hillel, 'Torque: The New Kinaesthetic of the Twentieth Century' in *Incorporations*, ed. Jonathan Crary and Sanford Kwinter (New York: Zone Books, 1992).

Schneider, Eckhard, ed., *Santiago Sierra, 300 Tons and Previous Works* (Cologne: Walther König, 2004).

Schulz, Bruno, *The Street of Crocodiles*, tr. Celina Wieniewska (Penguin: 1977).

Sebald, W. G., *Austerlitz*, tr. Anthea Bell (New York: Random House, 2001).

——, *The Emigrants*, tr. Michael Hulse (London: Vintage Classics, 2002).

——, *Rings of Saturn*, tr. Michael Hulse (London: Vintage Classics, 2002).

Serres, Michel, *The Birth of Physics*, tr. Jack Hawes (Manchester: Clinamen Press, 2000).

——, *The Five Senses: A Philosophy of Mingled Bodies*, tr. Margaret Sankey and Peter Cowley (London: Continuum International Publishing, 2008).

Seymour, Benedict, 'Eliminating Labour: Aesthetic Economy in Harun Farocki', *Mute Magazine* (14 April 2010) <http://www.metamute.org/editorial/articles/eliminating-labour-aesthetic-economy-harun-farocki>, accessed 4 April 2017.

Simoens, Tommy, ed., *Stan Douglas: Midcentury Studio* (Bruges: Ludion, 2011).

'Simon Starling: Recent History' (5 February–20 May 2011) <http://www.tate.org.uk/whats-on/tate-st-ives/exhibition/simon-starling-recent-history>, accessed 30 January 2017.

Soden, Garret, *Falling* (New York: W. W. Norton & Co., 2003).

Staus, Erwin, 'The Upright Posture', *Psychiatric Quarterly*, 26 (1952), 529–61.

Steinberg, Leo, *Other Criteria in Twentieth-Century Art* (Oxford: Oxford University Press, 1976).

Steiner, Shepherd, *Rodney Graham: Phonokinetoscope* (London: Afterall, Central St Martin's College of Art and Design, 2013).

Thater, Diana, 'A Man Becomes Unstuck in Time in the Film That Became a Classic!' in *Robert Smithson Spiral Jetty: True Fictions, False Realities*, ed. Lynne Cooke and Karen Kelly (Berkeley and Los Angeles: University of California Press, 2005), 168–70.

'Tight Roaring Circle', *Artangel* <http://www.artangel.org.uk/project/tight-roaring-circle/>, accessed 9 September 2013.

'Traces of Gravity' <http://whitecube.com/exhibitions/damin_ortega_traces_of_gravity_masons_yard_2012/>, date accessed, 30 January 2014.

Trainor, James, 'Martin Kersels', *Frieze*, 60 (2001) <https://frieze.com/article/martin-kersels>, accessed 4 April 2016.

Trotta, Roberto, *The Edge of Sky: All you need to know about the all there is* (New York: Basic Books, 2014).

Tschumi, Bernard, *Architecture and Disjunction* (Cambridge, MA: The MIT Press, 1996).

Vidler, Anthony, *Warped Space: Art, Architecture and Anxiety in Modern Culture* (Cambridge, MA: The MIT Press, 2000).

Walker, Stephen, 'Baffling Archaeology: The Strange Gravity of Gordon Matta-Clark's Experience-Optics', *Journal of Visual Culture*, 2 (2003) <http://www.sagepublications.com>, 161–85, accessed 18 February 2016.

——, *Gordon Matta-Clark: Art, Architecture and the Attack on Modernism* (London: I. B. Tauris, 2009).

Warner, Marina, *Richard Wentworth* (London: Thames and Hudson, 1993).

Weil, Simone, *Gravity & Grace*, tr. Emma Craufurd (London: Routledge, 1997).

Weller, Shane, 'Beckett and Ethics' in S. E. Gonstaski, *A Companion to Samuel Beckett* (Chichester: Wiley-Blackwell), 118–29.

Westerman, Frank, *Brother Mendel's Perfect Horse: Man and Beast in an Age of Human Warfare* (London: Vintage, 2013).

Williams, Raymond, 'Culture is Ordinary' (1958) in Ben Highmore, ed., *The Everyday Life Reader* (London: Routledge, 2002).

——, *Politics of Modernism: Against the New Conformists* (London: Verso, 2007).

Weyergraf, Clara, ed., *Richard Serra, Interviews 1970–1980* (New York: The Hudson River Museum, 1980).

Wood, Catherine, *Yvonne Rainer: The Mind is a Muscle* (London: Afterall Books, 2007).

Yass, Catherine, 'Keynote Speech', *Vertigo in the City: Conversations between the Sciences, Arts & Humanities*, University of Westminster, 29–30 May 2015.

Index

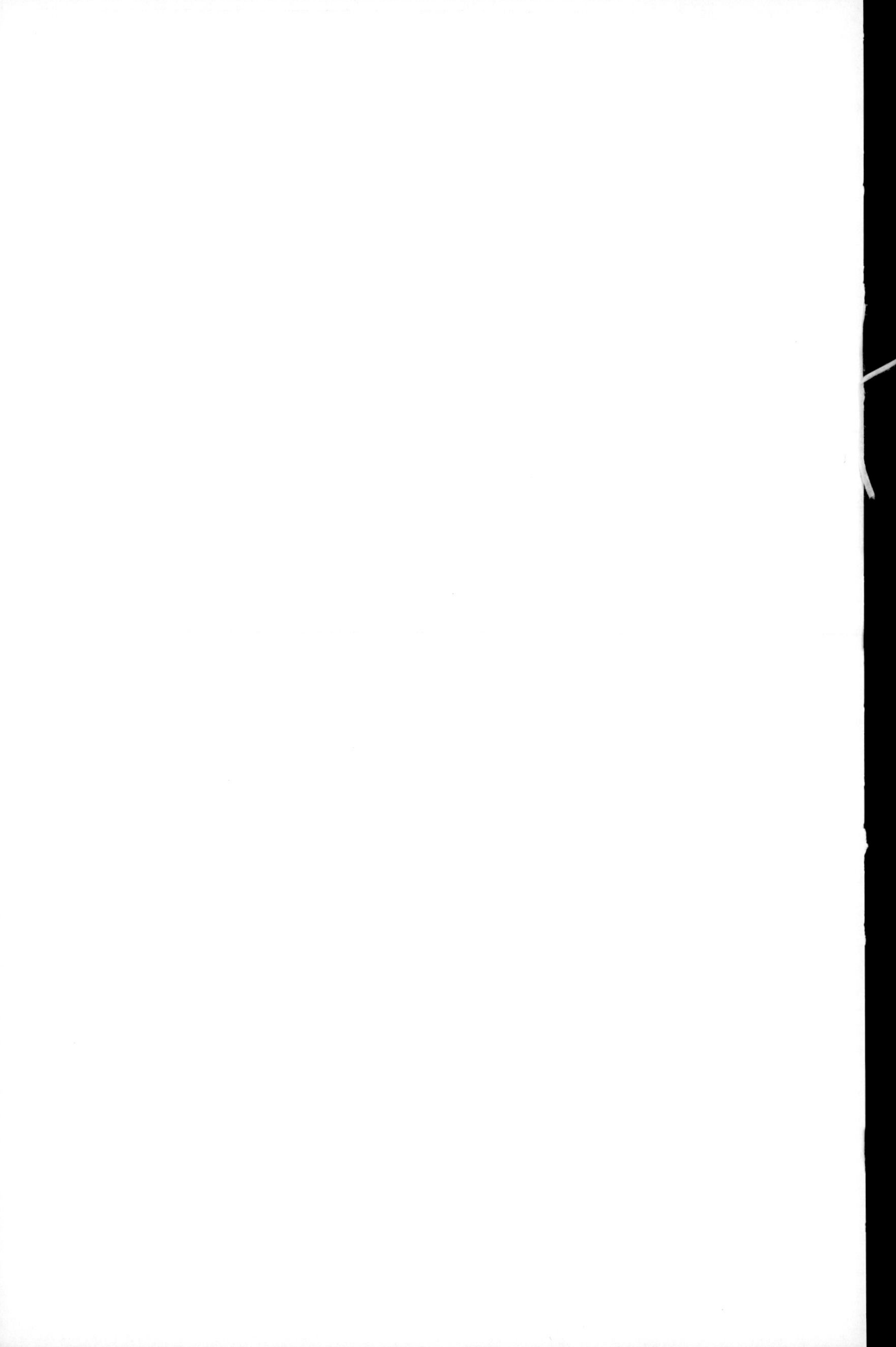